Information Society and the Workplace

In recent years there has been a great deal of talk about information society – in business, workplaces, organizations, management, government, academia, media and politics. This book looks at how the twin notions of information and society are entwined in workplaces and organizations.

Information Society and the Workplace brings together studies of everyday local practices in workplaces within the information society, with a special focus on social space and the agency of actors. The book makes use of both theoretical reviews and detailed qualitative research from Finland, which is at the forefront of the information society.

Key chapters include:

- Bounded or empowered by technology? Information system specialists' views on people's freedom within technology.
- The dynamics of control and commitment in IT firms.
- Women and technological pleasure at work?
- Fulfilment or slavery? The changing sense of self at work.

Highlighting the political challenges of the information society that are likely to become increasingly pressing worldwide, this book will be of essential interest to students and researchers in the fields of Business and Management Studies, Technology Studies and Sociology of Work and Organizations.

Tuula Heiskanen is Research Director of the Work Research Centre, a unit of multidisciplinary working life studies in the Research Institute of Social Sciences, University of Tampere, Finland. She has studied professional and industrial work, work organizations in both public and private sectors, and both male- and female-dominated organizations. She is co-editor of *Gendered Practices in Working Life* (Macmillan 1997).

Jeff Hearn is Research Professor in Sociology, University of Huddersfield, UK, and Academy Fellow and Professor at the Swedish School of Economics and Business Administration, Helsinki, Finland. His interest in information society comes particularly from research on gender, sexuality, violence and ICTs. His many books include *Gender, Sexuality and Violence in Organizations* (Sage 2001).

Routledge Studies in Technology, Work and Organizations

Series edited by David Preece

Information Society and the Workplace

Spaces, Boundaries and Agency

**Edited by Tuula Heiskanen
and Jeff Hearn**

Routledge
Taylor & Francis Group

LONDON AND NEW YORK

First published 2004
by Routledge
11 New Fetter Lane, London EC4P 4EE

Simultaneously published in the USA and Canada
by Routledge
29 West 35th Street, New York, NY 10001

Routledge is an imprint of the Taylor & Francis Group

Typeset in Baskerville by
Taylor & Francis Books Ltd
Printed and bound in Great Britain by
The Cromwell Press Ltd, Trowbridge

British Library Cataloguing in Publication Data
A catalogue record for this book is available from the British Library

Library of Congress Cataloging in Publication Data
A catalog record for this book has been requested.

ISBN 0–415–27223–8

Contents

PART III
Working and living in spaces

PART IV
Concluding discussion

Contributors

Pernilla Gripenberg is a doctoral researcher in the Department of Management and Organization, Swedish School of Economics and Business Administration in Helsinki. She is currently finalizing her thesis on interpretations of the information society, where she is studying ICT use in everyday life contexts, such as home, workplace and community. She has written and edited a series of reports in the IT_FAMILIES project. She has also published in *The Information Society* journal and in *Proceedings of the European Conference on Information Systems*. Her research interests include the social study of technology perspective, in particular for studying the human side of ICT including learning, computer literacy, digital divide, end users of ICT, virtual organizations and environments, and information society. She has also taught international business and business ethics.

Jeff Hearn is Research Professor in Sociology, University of Huddersfield, UK, and Academy Fellow and Professor, Swedish School of Economics and Business Administration, Helsinki, Finland. He has published very widely on work, organizations and management. His interest in information society comes particularly from research on gender, sexuality, violence and ICTs, including that with Wendy Parkin in *Gender, Sexuality and Violence in Organizations* (Sage, 2001), which examines ICTs, and globalization; that with Marjut Jyrkinen on the "Sexualized violence, global linkages and policy instruments" project, on policy development on the sex trade in the light of the growth of ICTs; and the Academy of Finland fellowship project on "Men, gender relations and transnational organizing, organizations and management".

Tuula Heiskanen is Research Director of the Work Research Centre (WRC), a unit of multidisciplinary working life studies in the Research Institute of Social Sciences, University of Tampere, Finland. She has a PhD in Psychology and has studied professional and industrial work, work organizations in both public and private sectors, and both male- and female-dominated organizations. She has a long-term interest in international scientific co-operation. With Liisa Rantalaiho, she co-edited *Gendered Practices in Working Life* (Macmillan 1997).

Riikka Kivimäki is a researcher in the Work Research Centre in the University of Tampere. Her main research interest has been to study the relationship of work and family. She has published articles on this theme, for example in L. Rantalaiho and T. Heiskanen (eds) *Gendered Practices in Working Life* (Macmillan, 1997), and a Nordic publication, J. Bonke (ed.) *Dilemmaet arbejdsliv-familieliv i Norden* (Nordisk Ministerråd, TemaNord 1997:534, Socialforskningsinstituttet 97:5).

Sirpa Kolehmainen is Assistant Professor in the Department of Social Policy and Social Work, University of Tampere. Her main research interests are employment and gender segregation in the labour market, gender equality in the working life, and knowledge-intensive work organizations. Her doctoral dissertation, *Gender Segregation on the Finnish Labour Market 1970–1990*, was published in 1999. She has taken part in the international comparative OECD project, "The future of female-dominated occupations", and participated in the national Finnish evaluation of the effects of the European Employment Strategy. Her publications include the OECD report *Occupational and Career Opportunities of Women in Female-dominated Occupations*, 1997, *Work Organisation in High-tech IT Firms* (Tampere University Press, 2001) and "Features of internal co-operation in high-tech project organisations" in E. Pantzar (ed.) *Perspectives on the Age of the Information Society* (Tampere University Press, 2002).

Päivi Korvajärvi is Professor of Women's Studies in the Department of Women's Studies, University of Tampere. She has extensive experience of ethnographic case studies both in public and private sector work organizations. Her recent publications include: "Gender-neutral gender and denial of difference", in B. Czarniawaska and H. Höpfl (eds) *Casting the Other* (Routledge, 2002); and "Locating gender neutrality in formal and informal aspects of organizational cultures", *Culture and Organization*,8 (2002).

Riitta Kuusinen is a Doctor of Psychology, working in the University of Tampere Work Research Centre. Her first career was in administrative and curatorial tasks in the Finnish and Swedish school systems. She then returned to the academic world to research the relationship between work and education. Studying the question of learning skills of co-operation at school and in working life has been a central concern to her. Her doctoral dissertation on knowledge-intensive co-operation was completed in spring 2001.

Riitta Lavikka completed her doctoral dissertation, *Big Sisters. Spacing Women Workers in the Clothing Industry. A Study of Flexible Production and Flexible Women in 1997* (Work Research Centre, University of Tampere, 1997). This is a long-term follow-up study of women workers' work culture during structural and work organizational changes in the Finnish clothing industry in the early 1990s. Following her first career as a journalist in labour and union journals, she returned to the academic world in 1991 to work as a researcher at the Work Research Centre, University of Tampere, on transformations of work organization in the clothing industry. Her main research interests lie in the sociology of work, organization studies and women's studies. She has also studied information society in the context of working life in large European and Finnish research programmes.

Tarja Tiainen is Acting Research Professor in Electronic Business at the University of Tampere. Her background is in information and computer sciences. She has studied the predispositions of the information system field and the roles of system specialists and technology users in shaping future technology. She has published in academic journals such as *AI & Society* and *Futures*, and at various information system conferences.

Marja Vehviläinen is a University Lecturer at the Department of Communication, University of Helsinki, and a Docent in Women's Studies and Information Technology, University of Jyväskylä, Finland. She was formerly Professor in Gender, Culture and Technology Studies, Luleå University of Technology, Sweden. She has conducted empirical research projects on the cultural and social construction of information technology, with a focus on gendering processes, in both the working life and broader information society contexts. She has published articles in international forums on gender, expertise and information technology as well as on local interpretations and agency in information society.

Preface
Paradoxes of information, society and workplaces

Jeff Hearn

Information. Society.

Information may suggest intelligence, knowledge, data, neutrality, transmission and communication: it is that which *informs* and is given meaning as *informative*. The term "information" may appear and be used ideologically as "neutral"[1] – declassed, degendered, without close reference to (any) social division.

Society may speak of agency, social order, partiality, division, collectivity and communication: it is that which provides the *social context*, which *socializes* and *gives meaning*. The term "society" may appear as and may imply social division – by class, gender and other social differentiations.

The combination of these two words, as in "information society", speaks to contemporary social conditions directly, dramatically, perhaps even aesthetically. The two words put together are much more than the sum of their parts. Together they convey, or seek to convey, some of the complexity of what might be called the current societal predicament.

Part of this predicament concerns how the twin notions of information and society are entwined together. This entwinement, and most specifically that in workplaces and organizations, is the subject of this book.

It may appear at first that this interconnection of information and society is simply part of the long history of the extension of calculability, instrumentality, rationality and rationalization. But, more accurately, it may have the quality of cultural construction, even undecidability, perhaps to the point of anti-foundationalism.Put together, within this perspective, "information" and "society" constitute and suggest a heady cultural mix. The notion of information society confronts grounded empirical, social change and the hermeneutic, interpretive nature of social life, including that around information.

There is clearly a great deal of talk nowadays of information society, in business, workplaces, organizations, management, government, academia, media, politics. The term has become a self-evident neologism, as in some societies the pervasiveness of information technology increases in many, probably most, spheres of human activity. Such societies may, after all, appear to pre-figure "things to come", even with the prospects of global catastrophe.

Having said that, it is important to remind ourselves that all societies entail some measure and flow of information and communication of information, even some form of "information technology". Information, like knowledge, is clearly not new to society: society depends upon information and communication of information for its production and reproduction, its continuity and change. It is very hard to imagine what a society might be without information of some kind. For social life to continue necessitates some form and process of informing others. And although information is given meaning within social and cultural processes, paradoxically through that it becomes an institutionalized subject giving meaning: information is thus that which informs.

However, in information society, information is not only about information in this general sense. Nor is information society even strictly about the use of knowledge, research and technical expertise within the production process. Rather, it is also about the complex and increasing infiltration of those forms of organized knowing beyond production in other areas of social life – consumption, family life, community, leisure, sexuality, sport, war and so on. Wang (1994) describes this as informatization:

> a process of change that features (a) the use of informatization and IT to such an extent that they become the dominant forces in commanding economic, political, social and cultural development; and (b) unprecedented growth in the speed, quantity, and popularity of information production and distribution.
>
> (p. 5)

While informatization moves beyond the worlds of production-based workplaces and organizations, it does so through the development of other workplaces and organizations.

In describing and analysing such societies, the term information society competes with other related terms, such as post-industrialism (Bell 1973), advanced capitalism (H. Schiller 1976), new economy (Carnoy 2000), informational capitalism, network society (Castells 2000), virtual society (Agres *et al.* 1998; Igbaria 1999; Shields 2002; Woolgar 2002), digital capitalism (D. Schiller 1999), digital economy, digital era, information economy, knowledge society, knowledge economy, postmodernity and postmodern society. Other permutations of key words can be imagined. More specifically, Schienstock *et al.* (1999) have distinguished between the following perspectives on the information society: information economy, post-industrial society, era of post-Fordism, informatized industrial society, knowledge society and learning society. There is thus now a huge set of different literatures on and relevant to information society.

In reviewing these literatures, Webster (1995, 2002) defines and analyses information society as a diverse process, from several perspectives: technological, economic, occupational, spatial, cultural. He also distinguishes two major schools of thought: those who see information society in terms of continuity, continuing pre-established social relations, broadly part of late modernity; and those who see it in terms of discontinuity, that a fundamentally new kind of society has emerged from the previous forms.

Largely following the former track, at least some of the features of information society can be traced back a very long way. Somewhat arbitrarily, perhaps paradoxically in these connections, one also could mention the part played by major individual theorists and practitioners such as Bacon, Saint-Simon, Comte, Weber and Nightingale[2] in the development of rationalistic ideas and ideologies. More broadly, there is the more general foregrounding of Reason from the Enlightenment onwards.[3]

Societal elements of information society are also not new, in the material sense that information work and information workers have been increasing since at least the mid-nineteenth century. This has occurred with industrialization, the development of the office, bureaucracy, management and the separation of ownership and control. It has been promoted through the growth of the modern state, applied science, gathering of national statistics and statistical analysis, technocracy, research and development, and the professions. More generally, the spread of mass education, mass media and electronic communication have all contributed fundamentally to the development of information society. Together, the information society can be seen as heavily interlinked to restructuring of economies away from primary and secondary sectors, and towards the tertiary and quarternary industrial sectors. Knowledge has been increasingly focused on knowledge; knowledge has been enacted to act on knowledge.

Most importantly, for the present purposes of this book, the notion of information society has specific implications for organizations and workplaces. The key importance of organizations and workplaces in the general societal process of rationalization has been highlighted by many commentators, though many have restricted their analysis to generalized macro-levels. In most of the various kinds of formulations of information society it is reasonable to assume that there are:

- increasing numbers and proportions of workers in organizations specializing in information production, handling, distribution, monitoring and use;
- increasing numbers of organizations specializing in information production, handling, distribution, monitoring and use;
- people and organizations increasingly affected by information production, handling, distribution, monitoring and use in more spheres of life.

A relatively early example of an emphasis on organizations and workplaces is Wilensky's (1967) study of *Organizational Intelligence: Knowledge and Policy in Government and Industry*. This pointed to the part played by complex organizations and states as the *collective action units* of modern society. It also showed the intense social constructions of information and intelligence, as well as ignorance and lack of or poor quality information and intelligence. While Wilensky's main focus was on the transition from organizational units of production of intelligence to political decision-makers and other elites, Gorz (1972) was more concerned with the *class relations* of different kinds of technical and scientific workers, and their "technical intelligence" within the capitalist production process. Gorz saw such workers as having both technical and ideological functions: that is, in relation to both the technical and the social relations of production.

By the mid-1980s, still in the pre-Internet days, Locksley (1986) analysed the intimate associations of information technology and capitalism, especially global capitalism. Faced with this information-driven political economy, he argued for a *political engagement, negotiation and confrontation* with information technology business and their controllers, particularly the largest multinationals in the field. Through this he recognized the possible democratizing opportunities provided by the "information explosion" throughout many spheres of society, so presaging many recent debates.

Increasingly, since then the argument has shifted to that whereby *the mode of information* begins to constitute society differently, in quite a different way than previously (Poster 1990). Thus the mode of or relation of information constitutes society more fundamentally. The form that this information-driven society takes is still derived from culture. Furthermore, information, the information itself, is always social, always in a social context. Recently, Lash (2002) has presented the perspective of informationcritique, developing but diverting from *ideologikritik*, developed within neo-Marxism to refer to the critique of dominant ideology, in its embrace of poststructuralism. Informationcritique addresses the changing social form of symbolic power from its being ideological to informational. According to Lash, this is essentially a matter of socio-cultural change towards global information culture, and thus the inherent contextuality of social action (Knorr 1979; Cooper and Fox 1990), not a question of philosophical absolutes.

What is being promoted in the more discontinuous concept of information society is that the historical relation of society to information operates in quite different ways. There is the notion of a particular kind of information that is present in the concept of the information society – information that appears disembodied, distant, abstract, and above all both unknown and trusted. Thus, paradoxically, it is the increasing use of information from sources that are *not known* that constitutes the particular kind of information that is the information of information society. In some so-called information societies, and probably increasingly so, the form, source, access to, verifiability and power of information lies *beyond* (the known).

Information and information society are thus neither transparent nor innocent. In this book a critical cultural approach to information, information society and information technology is adopted. To adopt such an approach is not to argue for simple *culturalist* explanations: that is, for a preference for explanation on the basis of or by way of autonomous "culture" over "economy" or "structure". Instead, culture here is a shorthand for the total social formation, including the economic and the structural. It is just that in some societies and some historical periods, certain interpretations of culture and the cultural appear dominant (see Lash and Urry 1994).

The ideas of culture and the cultural clearly have a very long history. Culture remains one of the most complex concepts in English and many other languages (Williams 1976). Many earlier approaches closely equated culture with "a people" (Levi-Strauss 1973), the "archipelago" vision of culture located in particular societies (Wright 1998). There is now, in contrast, a *growing diversity* in conceptualizations of culture. The idea of culture is increasingly contested: "it is … not neutral, it is historical, specific and ideological" (Swingewood 1977: 26).

Culture is used to talk of a very wide variety of social phenomena: social solidarity, class experience, movement from tradition to modernity, material change, and so on. Colonialist meanings of the "cultures" of "others" have been highlighted (Said 1978). Feminist critiques (e.g. Morris 1988) have gendered culture and constructions of women through culture; critiqued culture as unified and shared; and shown how the concept can obscure men's domination and construction of women, not least through associations of women with nature, and men with culture (Ortner 1974; Moore 1987). Some cultural critiques and critiques of culture have also pointed to the implicit and explicit construction of men in culture.

The notion of culture has also been used as a central feature of both late modern and postmodern society with its own realms of signification, consumer culture and semi-autonomous, multiple cultures (Foster 1985; Jameson 1991). Complex, contextualized uses of culture have been promoted, through the semi-autonomy of cultural regimes and new cultures set in contradictions of the local and the global. Above all, " 'culture' is not a 'thing' but a political process of contestation over the power to define concepts, including that of culture itself" (Wright 1998: 12).

To observe and figure the cultural is to engage with the contextualized, provisional, ambiguous nature of everyday life, including that of and in information society. This book sees information society in some societal milieux mirroring Grossberg's commentary on "popular culture" as "not … a fixed set of texts or practices, not … a coherent ideology … It is the complex and contradictory terrain, the multidimensional context, within which people live out their everyday lives" (1988: 25). This includes fundamentally changing forms of social time(s) and social space(s) (Meyrowitz 1985).

More precisely, in such a critical cultural perspective a number of key features of information in organizations and workplaces in "information society" are emphasized:

- First, information, *what counts as information*, is context-specific socially and societally, at work, in workplaces and in organizations. It is a socio-cultural, societal phenomenon and construction.
- Second, information is increasingly a *computer-mediated production and reproduction*. As such, it is only residually dependent on the hand; it is beyond the hand.
- Third, information is increasingly produced *from beyond the specific organization and organizational workplace*, as in a fundamental extension of the production

line. Organizations and organizational workplaces thus appear increasingly abstracted and disembodied (James 2001).

- Fourth, information is increasingly *from beyond society or a given society*. Thus, informationization is intimately linked to desocietalization (O'Brien and Penna 1998). This is not only a political process of fragmentation of nation-states, but a cultural process, brought by the transcending of boundaries by information and cultural artefacts.

So, there are a number of *paradoxes* around the relations of information to culture, technology, organization and society. Increasingly, in this kind of society both organization and society are formed in their transcendence. Workplaces and organizations are characterized by information that transcends the organization and the workplace. Information society involves an increasing focus within society on organizations and workplaces that are characterized by their own transcendence.

Thus this book focuses on case studies of organizations and workplaces where information is given meaning from beyond the organization: what might be called, admittedly with some difficulty, the process of *de-organizationization*! And these case studies analysed in this book are located within a society, Finland, where this process of desocietalization is particularly advanced, where information is given meaning from beyond the society.

Finland provides a very interesting context. This is a society with a very high level of use and development of information and communication technology, and high levels of higher education, international competitiveness, labour flexibility, and social and economic performance (Koistinen and Sengenberger 2002). The UN index of technological advancement (TAI) places Finland as the most technologically advanced country in the world. Finland has been top of International Data Corporation's information society index (ISI) since its inception in 1996, and has the highest number of Internet servers per capita, together with the US, and the highest level of mobile phone use (Castells and Himanen 2001; *Socially Responsible Information Society* 2001; Karvonen 2001). In 1999 Finland had the highest per capita of mobile phone (57.8/100) and Internet subscriptions (107/1000) in Europe; in 1998 nearly two-thirds of Finns had access to a computer, and in 1997 the same proportion of wage earners were using a computer in their work (Parjo 1999). The latest figures from the Gallup survey *Kids Online*(http://www.gallupweb.com/) indicate that over 70 per cent of 10-year-olds access the Internet from home. Finland is at the forefront of the development of an information society.

Accordingly, this book has four specific, and in some ways paradoxical, features, in:

- dealing with the everyday life of "information society", through *detailed, ethnographic and qualitative research*;
- addressing the *everyday, cultural, local, embodied experiences of and agentic practices around information technology*;

- focusing on those everyday, cultural experiences of information technology in *workplaces and organizations*; and
- examining these matters in *Finland*, the society that can most strongly claim to be or to be becoming an information society.

These paradoxical features are returned to at the end of the book.

Finally, in a book about knowledge and information, it is appropriate that we are also concerned with knowledge and information about knowledge and information. Studying information societies raises major questions of method and methodology. How are we to study such societies? Can rationalistic methodologies study what purport to be rationalistic phenomena? Are socio-technical methods needed to study socio-technical change? What are the limits of qualitative research in this area? What does a critical cultural approach look like? How are grounded accounts and explanations to be produced? How might comparative studies be developed across and between societies, languages and cultural forms? These questions preoccupy us and, we trust, others too.

Notes

1   Roszak wrote in 1986 (p. 19) (cited in Webster 2002, 23):

> Information smacks of safe neutrality; it is simple, helpful heaping up of unassailable facts. In that innocent guise, it is the perfect starting point for a technocratic political agenda that wants as little exposure for its objectives as possible. After all, what can anyone say against information?

2   As so often with "founding parents", these parents happen to be men. Lynn Macdonald (1997) provides a useful introduction to the often unacknowledged work of such women founders, including Florence Nightingale, who developed, following Adolphe Quetelet, what she called "the most important science in the world", social physics, a combination of societal analysis, philosophical rationalism and applied statistics.

3   In a telling critical evaluation of the relation of social theory and political practice, and specifically positivist social science and technological politics, Fay (1976) outlines the ideological presuppositions in the reduction of politics to technical solutions. He summarizes rationalization as: "the process by which growing areas of life are subjected to decisions made in accordance with technical rules for the choice between alternative strategies given some set of goals or values" (1976: 44), and continues

> the process of rationalisation pushes outward from its original economic base – where the development of natural science seemed most important – so that it finally includes politics as well – in which the development of a technically exploitable social science is thought to be an urgent necessity.

(pp. 46–7)

Acknowledgements

There are a number of people and organizations whom we would like to thank for their support in completing this book. The Academy of Finland gave financial support to the group via its Information Research Programme. Further funding has been received from EU Framework 5, ESF, the Finnish Ministry of Trade Technology Programme and SITRA, all of whom are gratefully acknowledged. The Work Research Centre, University of Tampere, has been the main site and scientific community for this work. The seminars organized by Gerd Schienstock with invited lecturers Sally Wyatt, Juliet Webster, Frank Webster, Ian Miles and Frieder Naschold gave the group an opportunity to learn from current theoretical discussions and ongoing research on information society in other countries. We are grateful to John Law for his inspiring seminar and comments on the chapter drafts. Liisa Rantalaiho gave us an exhilarating brainstorming environment to reflect on our research work in her beautiful lakeside summer cottage and in the Finnish sauna – the basic spiritual scene for Finns; and Liisa Husu gave support and advice. Finally, we want to thank our colleagues, Satu Honkala who helped with technical matters in the text production, Marjukka Virkajärvi who translated some parts of the text, Joan Löfgren who as a native English speaker has given valuable assistance in making our English more readable, and Hertta Niemi and Teemu Tallberg for bibliographic assistance and reading drafts of the beginning and end.

Part I

Introduction

1 Spaces, places and communities of practice

Tuula Heiskanen

Introduction

During periods of profound societal transitions, changes in work, employment and occupational structures are characteristically seen as an expression and indicator of systemic change. In the contemporary era, which has been described with a number of phrases such as the emergence of an information society, knowledge society or network society, we can observe substantial changes in work and labour organization as well as in organizational forms that structure work processes in new ways.

The increase of service jobs and decrease of manufacturing jobs has been a thorough-going tendency in advanced countries over many decades, albeit at a different rate in different countries (Castells 1996: 215; *World Employment Report* 2001: 5, 109). The path of development is clear but the interpretation of the kind of change behind the numbers is more ambiguous. Manufacturing jobs and service jobs as statistical categories have become less informative. The diversity of activities in the service jobs category has expanded and the dividing line between manufacturing and service jobs has become more blurred. Special attention in the development trends has been paid to the increase of the information content of work, and a special statistical category, "information occupations", has even been suggested (Porat 1998). Interest in the information content of work has been based on the notion in social theories that ongoing development is characterized by accelerating informatization of all spheres of social life and the increasing importance of the ability to use and generate knowledge, both in production and in society at large.

The desire to capture the dominant trend of development via the identification of information occupations and their share of employment has run into a definitional problem: what to include and what to leave out. Also, discussion around the term "knowledge worker" has suffered from the same problem, even though the scope of the concept is more confined, referring mainly to expert tasks which require creativity and innovativeness (Blackler 1995; Cortada 1998; Davenport and Prusak 1998; Alvesson 2001; Blom *et al.* 2001). The main difficulty in definition has been the lack of knowledge on the real content and contexts of work. This lack of an empirical basis has also prevented seriously tackling the core question of what information or what knowledge counts as a definitional criterion, a question that can only be answered in relation to the process of doing work activities.

This book focuses on *doing* and *being* at work. With this aim, important information about doing and being at work includes, besides occupational positions, types of employment contracts, organizational forms in which work takes place, and the role of technology. All these conditions have undergone substantial transformations during recent years. Statistics from the OECD countries and the European Union have shown an increasing tendency towards an increase in the proportion of temporary and part-time workers (*World Employment Report* 2001: 17). These indications belong to a picture of reduced security and predictability in employment relationships. There are also signs of individualization of work, which has been pushed forward by management practices through which management tries to reduce long-term employment relationships and which have resulted in a decrease in employees' loyalty to a specific organization. Indeed, instead of increasing this loyalty, they have produced greater attachment to personal careers (Cappelli 1999: 1–2; Carnoy 2000: 3–5).

"Flexibility" is nowadays often the guiding principle in business strategies in striving to adapt to market conditions. Part-time and temporary work and other new types of employment contracts provide labour flexibility, while new organizational forms provide flexibility in work processes. The need for tightening co-operation in rapidly changing situations has resulted in new organizational forms, from team and project organizations to virtual organizations which cross traditional boundaries of organizational structures. The integration of production processes has been supported by information and communication technologies, which have helped to cope with information flows and to overcome constraints related to time and space.

In writings on information society or the knowledge society, much emphasis has been placed on the role of information or knowledge in society. Castells (1996: 17, 21) has used the concept of the "informational paradigm" to stress not only the important role of information and knowledge but the emergence of "social organization in which information generation, processing and transmission become the fundamental sources of productivity and power because of new technological conditions emerging in this historical period". The informational paradigm opens up visions and challenges for alternative ways of organizing work processes. The positive perspectives in the visions are still far from realization in the actual forms of flexibility.

This book presents cases from traditional and new types of organizations and organizational settings which give a view on manufacturing and service jobs, IT-expertise, project work and telework. New information and communication technology plays an important but different role in each of the cases. In some cases, the technology is a prerequisite for the very form of the activities; in some others it is an element which gives new possibilities for the organization of the work process.

Cornfield *et al.* (2001: ix–x) have argued that workplace and workplace community give a contradictory starting point for social scientific studies, due to transformations in employment relationships and work arrangements. This book focuses, instead, on spaces of various kinds. The increase in the intensity of connections between economic actors and demands of complex ways of using knowledge at every level of work organization have transformed networks and relations between actors, giving rise to new kinds of social and cultural spaces. The book approaches the spaces through an agency perspective, which in its broadest sense generates the question of whether the actor has an influence on the events around himself/herself, and, if so, what kind of influence. The book asks: how is it possible to create spaces that satisfy multifaceted knowledge needs in working life? How is it possible to create spaces that increase rather than restrict human agency at work? What is the role of technology in these new working spaces? How are spaces and communities of practice related? What do the changes mean for changing forms of competences? And what kinds of impacts do the changes have on individual selves, identities and gender orders?

The cases come from Finland, a country with a Nordic-type welfare state and a highly advanced technological infrastructure. In and through the cases it is possible to read expressions of the informational paradigm in doing and being at work.

Information Society

For the contributors to the book, discussions around "information society" have been a key context within which the case studies have been reflected upon and analysed. In contrast to a large number of macro-level writings, this book focuses on local practices. The information society discussion has dealt with the increase of information in different fields of society, flows of information, the connectedness of actors in different parts of the world, the shrinking of time and space, changes of production and occupational structures, and the role of information and communication technology in shaping practices and processes in society. These are questions which this book touches on, in the specified context of working life and as seen through the eyes of the multiple actors of working life.

Discussion on the information society has taken place in two waves. Daniel Bell's (1973) epochal book *The Coming of Post-industrial Society* paved the way for the first wave in the 1970s. Both concepts of the information society and knowledge society (Drucker 1969) were used in the discussion, which: forecast a move from the production of goods towards the production of services, and resulting changes in occupational structures; forecast the rise of the information economy and an increased role for knowledge; and touched upon questions of technology and the global order of the economy. The second wave of discussion starting in the 1990s has been backed by governments' and international organizations' increased interest in the potential of the information society. The idea of the information society has become part of the strategic thinking around development in the industrialized countries, as reflected in strategy documents in the US, for example, in the National Information Infrastructure Programme, 1993, in the European Union *White Paper on Growth, Competitiveness and Employment* (1993), and similar documents in many other countries.

It is characteristic of present-day discussions on information society that they include a mixture of political talk, utopian talk and talk by social scientists. Politicians' interest in the information society has guaranteed wide publicity for the discussion. Utopian discussion has given wing to visionary speculations in a variety of forums. The role of technology in future developments arouses passions, with Toffler (1980) and Masuda (1980) often cited as authorities. Among social scientists, there have been reservations whether the concept of the information society, or knowledge society, is a valid expression. Webster (1995, 2002), for example, has analytically distinguished five definitions of the information society which proponents of the idea have used. Developments diverging from the criteria embedded in the definitions do not convince Webster that the information society has been attained.

The most often-used definition of information society is based on *technological* development. It is a plain fact that the application of information technologies has spread to virtually all fields of society; in particular, the convergence and imbrication of telecommunications and computing has opened the channel for links within and between offices, banks, homes, shops, factories and schools. The question which Webster poses is: how much IT is required in order to identify an information society? He has not found any satisfactory answer in those writings which use a technological definition.

Another popular definition of information society looks at the share of the information sector of the whole *economic* activity of a country. Webster's scepticism towards this kind of information economy claim comes mainly from measurement problems. How to categorize informational and non-informational domains? Which economically assessed characteristics are more central to the emergence of an information society? These are questions which, according to Webster, have not been answered adequately yet but which need to be settled if we are to arrive at reliable conclusions about the predominance of the information sector.

Another perspective used to look at economic activities has been *occupational* change. According to this definition we have attained an information society when the predominance of occupations is found in information work. This definition runs into parallel problems to those found in the studies focusing on distinguishing between the information sector and other economic functions.

The *spatial* definition of information society emphasizes the information networks that connect locations and have substantial effects on the organization of time and space. Webster agrees that information networks are an important feature of contemporary societies but challenges commentators to specify why information networks, on which there has been a long-term dependency, comprise a qualifying feature for a new type of society just now.

As the fifth type of definition, Webster mentions the *cultural* characterization of everyday lives. From a cultural perspective it is easy to see the media-saturated environment in which we live, as well as the expansion of the informational content of everyday life and social intercourse. As compared to the other definitions, this conception of the information society is least measured and examined.

Behind the different perspectives on information society, the overarching theme in the writings is the emphasis on the key importance of information to the modern world. Of course, knowledge and information have played an important role in all societies. Subscribers to the notion of information society claim, however, that the significance of information differs from that in hitherto existing societies. Castells (1998: 338–9) presents this argument by saying that in the information mode of development, both culture and technology depend on the ability of knowledge and information to act upon knowledge and information, in a recurrent network of globally connected exchanges. In explaining this assertion (1996: 17) he says that informationalism is oriented towards technological development: that is, towards the accumulation of knowledge and towards higher levels of complexity in information processing. Considering the salient role of information in the present era, Lash (2002: 2) suggests focusing on the primary qualities of information itself, on flow, disembeddedness, spatial compression, temporal compression and real-time relations when trying to understand the information society. Lash argues that the term "information society" is preferable, for example, to "postmodernism" simply because the former says what the society's principle is.

Finland provides a special context in which to study issues related to the information society. On a number of measures Finland is in the forefront of the development of an information society. Castells (2000: 72) suggests:

> The Finns have quietly established themselves as the first true information society, with one website per person, internet access in 100 per cent of schools, a computer literacy campaign for adults, the largest diffusion of computer power and mobile telephony in the world, and a globally competitive information technology industry, spearheaded by Nokia.

In international comparisons, new information technology applications such as Internet use and the spread of mobile phones are often cited measures. On such lists Finland and other Nordic countries are ranked highest (*World Employment Report* 2001: 335–49). The true significance of Finland's experiences in the information society context is not, however, simply based on the extent to which certain types of technology are used. More interesting is the path of development that has resulted, among other things, in the advanced technological infrastructure, which is concretely visible in the statistical figures. The development of Finland up to the present "stage" of "information society" has taken shape as a consequence of several interacting factors. Both long-term and short-term processes can be found behind the development, such as transformations in industrial structure, technology policy, labour market relations and cultural traditions in the face of crises (cf. Kasvio 2000).

Finland is a small and relatively open economy, and developments in the global economy are strongly felt throughout the whole society. A recent experience from the beginning of the 1990s provides a good example: the consequences of the worldwide economic depression were exaggerated in

Finland by the simultaneous collapse of the former Soviet Union. As a neighbouring country, Finland had extensive bilateral trade with the Soviet Union. When the trade suddenly reduced to a fraction of the former level, Finnish companies were forced to find new markets for their products. This new challenge hit some labour-intensive branches especially severely (for example, the textile and clothing industry, the construction industry). The unemployment rate rose within a short time from 3 per cent to about 20 per cent. The depth of the economic depression was a national crisis, which demanded a wide consensus in seeking paths leading out of it. The labour market bargaining system, with the government involvement in the frame-negotiations, ensured that the labour market parties also shared the crisis consciousness. Putting it in broad terms, the solution was twofold: to develop the industrial structure and related infrastructure, and to save the Nordic-type welfare state.

The Finnish economy has traditionally been heavily dependent on the country's rich wood resources and wood-related industry. The Finnish paper industry, for example, continues to be a big player globally. The industrial structure, however, changed substantially in the 1990s. Now the wood and paper industry is only one of the three major export sectors, the other two being electronics and metal and engineering (Statistics Finland 2002). Electronics is the most spectacular success story in recent Finnish industrial activities, mainly based on mobile phones and telecommunication equipment.

The successful orientation to the world market after the early 1990s crisis was not just a lucky coincidence. It was preceded by a number of supportive actions. Developments in science, technology and innovation policy, in education policy and in research and development (R&D) investments are noteworthy in this regard. Throughout the 1980s many discussions within Finnish industry raised the need to seek competitive advantage from knowledge intensity and technological development rather than from cost efficiency as such. The discussions concretized in the rapid increase in R&D investment. Science, technology and innovation policy shared the same objectives. Finland was the first country among the OECD countries to adopt a systemic approach on such policy by using the concept of a national innovation system as a framework (Ormala 1999). Other OECD countries have made similar kinds of efforts in adjusting their policy and seeking an adequate role for governments in facilitating and providing infrastructure (OECD 1997). On a practical level, it has meant an attempt to deal with all the elements that contribute to the generation, diffusion and application of new knowledge simultaneously within the same framework. Educational policy has involved a dialogic relationship to national innovation policy, and the educational system has been able to provide the competences needed in the knowledge-intensive high-growth areas.

Technological development as the factor which opens future prospects has wide support among Finns (EVA 1999). National strategies for the information society in different countries have a strong technological element, especially in relation to advances in information and communication technologies. The same applies to

Finland. The first strategy document (*Suomi tietoyhteiskunnaksi* 1995) was formulated by a committee originally commissioned to deal with technological concerns. After the publication of the US strategy and the start of discussion in the EU, the committee autonomously broadened its tasks. After this, the continuation of national strategy has been active and dynamic. Different kinds of actors have been involved in this work at both national and local levels, and the significance of the strategies is widely acknowledged in the public discussion. In addition to the technological concerns, the discussion has also moved further towards social concerns.

Balancing between technological and social orientations is of major concern throughout the world on many levels and among different kinds of actors, including academics, industrialists and policy-makers. As a challenge to the dominant discussion, Kasvio (2000) thinks it would be worth considering the benefits of an egalitarian social structure and democratic education, as a counterpoint to a market-driven transformation process which seems to divide citizens sharply into winners and losers. This kind of comparison is also in Castells' mind when he presents Finland as an example of an information society from which other countries might learn. The earlier quotation from Castells (2000: 72) continues: "At the same time they have kept in place, with some fine-tuning, the welfare state." Thus he notes that, along with the plans and actions striving for technological advance, Finnish people have been committed to the welfare state and related values of social solidarity, even during the years of economic crisis and under pressure for substantial reduction in public expenditure. This information society strategy, emphasizing both technology and the welfare state, has become in the minds of Finnish people a kind of identity project which has an imprint of survival over difficulties and at the same time a strong future orientation (Castells and Himanen 2001).

The book focuses on the daily life of organizations and their actors, taking macro-level policies and orientations as its contextual framework. Macro-level strategies and policies define conditions for agency, on the one hand, by providing the means and producing competences for access to the technology and structures of the information society (Lash 1995: 182–3; Wyatt *et al.* 2000). On the other hand, they also define conditions for agency by shaping general orientations on how surrounding circumstances can be influenced. The cases open up a perspective on learning and problem-solving situations, which nowadays are often framed as problems of knowledge management and as challenges in providing the workforce with competences which are increasingly social in nature. The cases are from both traditional and new kinds of organizations, each of which continue to function in the "information society". The conditions for action described in the case studies presented are specific to the organizational settings, but the identities of the actors and their experiences of action competences are rooted in the wider context.

Spaces

Space is the research focus of many disciplines. It has become a major concern of social theory relatively recently, due to the emergence of new forms and processes of cultural and social spaces. The questions that have been highlighted relate to both physical/material and social/cognitive/mental aspects of space. Teamwork, project-based work, telework and work crossing organizational boundaries locally or globally are all examples from working life which have needed new kinds of thinking: how to organize relations of actors in ways which are most appropriate for the activities, what kind of material support is needed for the network of relations in work organizations, and how people cope with the new social settings.

Manuel Castells (1996: 411–12) has introduced "space of flows" as one of his key concepts in the analysis of societies in the information age. He argues that contemporary society is constructed around flows: flows of capital, flows of information, flows of technology, flows of organizational interaction, flows of images, sounds and symbols. He thinks that flows are the expression of processes that currently dominate economic, political and symbolic life. From this line of thinking follows his hypothesis of the emergence of a new kind of space which supports these dominant processes in our societies. He defines the space of flows as "the material organization of time-sharing social practices that work through flows". By "time-sharing social practices" he means that space brings together those social practices that are simultaneous in time. The definition aims to account also for material supports of simultaneity that do not rely on physical contiguity. In the information age such social practices are increasingly important.

The concept of space has multiple meaning: for example, in physics and geometry as against sociocultural phenomena. Sorokin (1943: 122) argues that in order to be adequate, the concept of sociocultural space must be able to define the position of any sociocultural phenomenon among other sociocultural phenomena or within the sociocultural universe.

Castells (1996) derives his concept of space from the analysis of the informational mode of economic activities. According to him, transformations in time and space take place under the combined effect of the information technology paradigm and social forms and processes induced by the current process of historical change. Giddens, another influential promoter of the space theme in social theory, grounds his ideas on space on the comparison of premodern and modern societies. He has emphasized that notions of time and space have to be brought into the core concerns of social theory (Giddens 1989: 275). Giddens (1990) considers that the examination of recombinations of time and space, separation of time from place and space from place, is crucial to understanding the dynamism of modernity. Accordingly, time and space refer to the contextualities of social interaction, or to the intermingling of presences and absences in the conduct of social life (1989: 276).

Characteristic to modernity has been the stretching of social relations across time and space, a process which Giddens calls "time–space distanciation". In premodern cultures time was linked with place, "when" was connected with "where". In addition, space and place largely coincided, since spatial dimensions of social life were dominated by localized activities. The emergence of the uniform dimension of "empty" time, separated from place/space, coincided, according to Giddens, with the expansion of modernity. The advent of modernity also started increasingly to tear space away from place by fostering relations between "absent" others, locationally distant from situations of face-to-face interaction.

Giddens' work has stimulated much discussion on how to build time and space into theoretical thinking about the nature of social life and social systems. One critical comment, which is of special importance for the approach adopted in this book, is made by Gregory (1989: 187). He claims that Giddens' theory remains close to the analytics of spatial science and has little to say about meanings of place and symbolic landscapes. Furthermore, he claims that Giddens is virtually silent about the "production of space".

Lefebvre (1998), in particular, has devoted his attention to the production of space. In everyday parlance, space is used in a variety of ways, for example to refer to specialized spaces such as leisure, work, play, transportation, public facilities. Spatial disciplines like geography, architecture, urban planning and urban sociology each provide their own further definitions to space. Lefebvre has been dissatisfied with this splitting of knowledge and has instead striven for a unified theory of space, unified in the sense that it encompasses physical, mental and social aspects of space. As an analytical tool he has suggested a conceptual triad, three moments of space, which he argues should be looked at simultaneously: perceived, conceived and lived spaces; or, in spatial terms, spatial practice, representations of space, and representational spaces (or spaces of representation, a term suggested by Soja (1996) instead of the English translation "representational spaces").

Spatial practice is defined as producing a spatiality that:

> embraces production and reproduction and the particular locations and spatial sets characteristic of each social formation. Spatial practice ensures continuity and some degree of cohesion. In terms of social space, and of each member of a given society's relationship to that space, this cohesion implies guaranteed level of competence and a specific level of performance.
>
> (Lefebvre 1998: 33)

Lefebvre goes on to argue that the "spatial practice of a society secretes that society's space; it propounds and presupposes it" (p. 38). And further:

> Spatial practice is that aspect of space which is empirically observable: it is observed, described, analysed on a wide range of levels, in architecture, in city planning, in the actual design of routes and localities, in the organization of everyday life.
>
> (p. 413)

A *representation of space* defines a "conceptualized space, the space of scientists, planners, urbanists, technocratic subdividers and social engineers, as of a certain type of artist with a scientific bent – all of whom identify what is lived and what is perceived with what is conceived" (p. 38). Representations of space are abstract but they also play a part in social and political practice (p. 41).

Spaces of representation embody "complex symbolisms, sometimes coded, some-times not". They are "linked to the clandestine or underground side of social life and also to art" (p. 33). This is a space which is "directly lived through its associ-ated images and symbols, and hence the space of inhabitants and users, but also of some artists, writers and philosophers, who describe and aspire to do no more than describe" (p. 39). While Lefebvre described conceived space as dominant, lived space is dominated and hence is passively experienced space, which the imagination seeks to change and appropriate. It overlays physical space, making symbolic use of its objects.

Lefebvre's approach takes as its starting point the realities of the present: the forward leap of productive forces, and the new technical and scientific capacity to transform natural space so radically that it threatens nature itself (p. 65). His conceptual work is rooted in an analysis of the dynamic of social life as seen from the perspective of the contradiction-filled interplay between the develop-ment of the forces of production and the social relations that are organized around production, and their continued renewal. From this basis he follows through his striving to link historicality, sociality and spatiality in a strategically balanced way, and from this basis comes also his hypothesis that "every mode of production produces a space, its own space" (pp. 31, 46). In this expression space should be understood in the meaning of dominant space and not as a denial of particular, different kinds of spaces.

In bringing together separate fields of knowledge concerning physical, mental and social space, he introduces the concept of "production of space", both as a theoretical concept and as a practical reality. For him (social) space is a (social) product (p. 26) and the space thus produced serves as a tool of thought and action. He makes an assumption that spatial practice, representations of space and spaces of representation contribute in different ways to the production of space, each according to the society or mode of production in question, and according to the historical period (p. 46).

Lefebvre's call to look at the dimensions of the physical, mental and social, or perceived, conceived and lived, simultaneously is not easily put into operational terms. Soja (1996) compares Lefebvre's solution to a musical composition with a multiplicity of instruments and voices playing together at the same time (p. 9), and more specifically to a polyphonic fugue based on distinct themes which are harmonized through counterpoint and introduced over and over again in different ways through the use of various contrapuntal devices (p. 58). In fact, Lefebvre uses the term "social space" in two ways, as Soja notes. It is, for him, both a separable field distinguishable from physical and mental space and also an approximation of an all-encompassing mode of spatial thinking, a tran-scending composite of all space. Relations between the three moments of the perceived, the conceived and the lived are never simple or stable in Lefebvre's descriptions.

Soja has further developed Lefebvre's conceptual triad on space, making it more comprehensible at the empirical level and as a research strategy. Soja (1996) introduces a new term, "Thirdspace", in dealing with the problems of multifaceted inclusiveness and simultaneities of social space (p. 58). According to him, the mainstream imagination has revolved primarily around a dual mode of thinking; one perspective is fixed mainly on the concrete materiality of spatial forms, on things that can be empirically mapped, and the second is fixed on thoughtful representations of human spatiality in mental and cognitive forms. Soja (1996: 10) claims that these coincide more or less with Lefebvre's perceived and conceived spaces, with the first often thought of as "real" and the second as "imagined". Lived space, which according to Soja is the preference implied by Lefebvre (Lefebvre does not explicitly place any preferences between the three moments of space), as it is by Soja himself, has not received the attention it deserves in research practices. The definition given by Lefebvre of lived space (spaces of representation) "lived through images and symbols" might give a misleading interpretation if one forgets Lefebvre's characterizations of these as spaces vitally filled with politics and ideology, with the real and imagined intertwined. Soja has taken the latter characterization as his main emphasis and stresses lived spaces as chosen spaces for struggle, liberation and emancipation (p. 68), as spaces of resistance to the dominant order. His primary interest is in fully lived spaces, which are simultaneously real and imagined, actual and virtual, in the lifeworld of individual and collective experience and agency (Soja 2000: 11, 351). He calls his focus Thirdspace, since it gives, according to him, an alternative third perspective that draws upon the material and mental spaces of the traditional dualism but extends well beyond them in scope, substance and meaning (Soja 1996: 11).

While Castells talks about the construction of society around flows and Lefebvre poses the hypothesis "every mode of production produces its own space", Wise (1997) starts his book by talking about a variety of spaces. These differences in emphasis (space/spaces) are a natural consequence of the starting points of the writers. While Castells starts from the informational mode of economic activities and Lefebvre from the interplay between the development of forces of production and social relations, Wise takes as his starting point the notion of agency. From this episteme, social spaces seem open and permeable spaces which are created through the interaction of multiple humans over time. Wise's understanding of agency is partly based on actor-network theory (ANT) (1997: xv) which examines situations in terms of the social effectivity of actors. Both humans and non-humans shape our social life and both are considered in this episteme as actors. Wise defines social space as a network of relations between actors (1997: 70). Wise is especially interested in how technology and social space intertwine, interact and are mutually constitutive. In defining social space, he states that

> social space is not just discursive patterns of "imaged", "imagined" or symbolic communities ... but neither is it only physical aggregates of individuals and constructed space ... social space is the stratification of the two and can be described as a series of actor-networks ... Social space always

consists of Technology and Language in particular configurations. But it also means something in addition to this configuration. It also means the embodiment of that configuration.

(1997: 70)

In Wise's usage, Technology and Language refer to two kinds of agency: Technology refers to corporeal agency, the ability to achieve effects through physical contact, and Language refers to incorporeal agency, the ability to achieve effects without corporeal means.

Wise argues that human social space is composed of the stratification of technology and language. Important questions to him are: what is the relationship between the two types of agency? What kinds of shifts occur between them in the surroundings of new technology? And what does it mean for human agency? In empirical investigations of social space, he considers Lefebvre's three perspectives on space to be valuable. Wise thinks that if social space is an actor network, it would consist of practices that make it up, renew it and transform it, as well as concepts of network itself and the representational experience of the network.

This book has been written in the spirit of the view that there is an inherent spatiality to social life. The book has been influenced in its spatial sensitivity by a number of writings on social space. Examples of time–space zoning given by Giddens indicate how to look at temporal and spatial organization of daily practices and routines. Urry's (1985) analysis of spatial divisions of labour also gives illuminative examples how to conceptualize social relations with the understanding that such relations are both temporally and spatially structured in a number of ways. Castells has made visible the global ties between economic activities in the formation of spaces. Lefebvre's and Soja's work has shown routes to comprehend physical, mental and social aspects of space in a unified way. Wise, with his examples, has illustrated the intertwining of technology and social space. The cases in this book represent examples of such temporal and spatial organization of activities which have gained importance in the development towards the information society (for example, telework, crossing organizational boundaries, virtual organization). Our interest is not, however, in spaces as such, as given arenas for social life, but rather in the lived world of actors in working life, within an appreciation of spatial sensitivity.

Places

In spite of the emergence of new forms of spaces (e.g. virtual spaces, spaces of flows) people's living experiences are overwhelmingly place-based. Still, the separation of space from place has given new meanings to both space and place. Castells (1996) has differentiated between the space of flows and the space of places. For him, a place is a locale whose form, function and meaning are self-contained within the boundaries of physical contiguity (1996: 423). Castells emphasizes the differentiation between spaces of flow and spaces of place based on his hypothesis that function and power in our societies are organized in the space of flows (1996: 428). He thinks that the structural domination of its logic essentially alters the meaning and dynamic of places. He is concerned about the break between these two spatial logics, as a consequence of which experience, by being related to places, becomes abstracted from power, and meaning is increasingly separated from knowledge. Soja (1996: 215) sympathizes with Castells' political intent to mobilize progressive political power of the space of places, but he rejects the dichotomized and totalizing conceptualization of the space of flows versus the space of places. Soja argues that there are examples of movements and practices that recombine abstract flows and concrete places and which are opening up new and different real-and-imagined spatialities of resistance and contention.

As an analysis of the tension between place-based and global activities, Castells' statements are insightful, but from the perspective of the lived world of actors we join in Soja's critique of the problems with the dichotomy. Wise (1997: 124) has defined the relationship between place and space by saying that space is a practised place, a place with actors. This view does not draw a sharp distinction between place and space; according to it, activity actualizes potentialities of place and thus creates social space. This line of thinking is congruent with the approach of this book, in which the understanding of space also includes meanings attached to places.

Actors and technology

Technology and, especially at present, information and communication technology (ICT) are inseparably part of the changes that can be observed in employment structures, organizational forms and workplace activities. In order to understand these ongoing changes, we need to keep technology in mind. The difficulty, pointed out by Wise (1997) in analysing the role of ICT is that it seems to be everywhere and still it disappears out of sight and out of the analysis. Wise himself uses actor-network theory in order to take a stance on technology. This book, too, utilizes perspectives that the spatial interpretation of technology suggests for a research orientation.

ANT grew out of the academic field of studies of science, technology and society. It has inspired a large number of empirical studies which have illuminated the goals, values, meanings, histories and social interests related to technology (Bijker *et al.*1987; Bijker and Law 1992). When ANT poses questions about technology it always does so in relation to empirically describable networks of actors. Actor networks, which are necessary for the functioning of societies and which are constantly established and reproduced, involve both human and non-human (among other things, technology) entities. The theory does not set out any *a priori* distinction between the human and non-human, or the social and technical in the construction and maintenance of networks. With that methodological orientation, the approach challenges both technological and social determinism.

Technological determinism in its strongest form presents technology as an autonomous force that determines social and cultural development. Social determinism stands in contrast to assumptions of technological determinism and presents technology itself as neutral and prone to social and political shaping. Even though such classifications of thought models are commonplace in polemic writings, it is easier to find a variety of interminglings than archetypal representatives of them (e.g. Grint and Woolgar 1997; McLoughlin 1999).

In contrast to approaches which struggle with the problem of dualism (technical or social) ANT strictly follows a symmetry principle. Callon (1986: 200) expresses it by saying: "The rule which we must respect is not to change registers when we move from the technical to the social aspects of the problem studied." The principle extends even to the definition of actors. Not only humans are actors but "any element which bends space around itself, makes other elements dependent upon itself and translates their will into a language of its own" (Callon and Latour 1981: 286). Or in Moser's (1998) words: "Human actors are not always subjects and things are not always objects." The question of agency is looked at focusing on results and effects. The well-known example given by Latour (1988) is that of an automatic door-closer which regulates entering into the room. Through its functions it manages space by regulating movement and human behaviour in that space.

Definitions of actors and networks are tied to each other. The interest in the analysis is not in the actor per se but rather in the actor's relations with other actors in network building and maintenance. In this sense, the actor and the network are two faces of the same phenomenon. ANT is a method of learning from the actors without imposing on them an *a priori* definition of their world-building capacities (Latour 1999: 20).

The proponents of the approach have been concerned about the confusions which might arise nowadays when terms like "World Wide Web" and "network organizations" are in common usage. In ANT the choice of the term "network" was originally meant to capture the contingent and emergent form in contrast to such notions as institution, society and nation-state. ANT draws our attention to the manner in which networks are built rather than any given social and technical entities, which is often the case in the present-day discussion on networks.

The symmetry principle in relation to humans and non-humans as well as to micro- and macro-actors has prompted several critical questions even though, on the other hand, its heuristic value is widely recognized. It has been asked, for example, whether insistence upon a symmetry between humans and non-humans means conceding to technical determinist accounts (Collins and Yearley 1992). It has also been asked whether looking at micro- and macro-actors according to the symmetry principle would lead to ignorance of real differences in power (Wise 1997: 35). Callon and Latour's (1992) response to this kind of critique is that their intention is not to prove symmetry but to use it as a methodological heuristic tool and to ask how and why asymmetries are established. ANT argues heavily against technological determinism, but Grint and Woolgar (1997) are not convinced that it has been successful in overcoming a form of technicism. They see instances in the writings of the founders of the theory in which the capacity of technology is treated as given, objective and unproblematic. Grint and Woolgar call this kind of assumption "technicism" (p. 31). They think that criticism of residual technicism gains particular strength from the presumption made by ANT that the aim of the social analyses of technology is to provide a causal explanation of technological development. If the aim, instead of being explanation, were to be a redescription, Grint and Woolgar consider that the character of the network's components would be less pressing.

Star (1991), Wise (1997: 34) and Winner (1993) have all criticized ANT for its tendency to fall back into dominant structures of power. The focus on the activities of network-builders leaves marginalized people outside the analysis, as Leigh Star has remarked, and the space that is described is, according to Wise, that of established power rather than the forms of resistance within that space. Winner (1993: 441) argues that the accounts of politics and society given by social scientists are implicitly conservative if they do not succeed in indicating which social groups have been sandbagged out of the laboratories and which social voices are being effectively excluded and silenced.

Much of the critique relates more to the limitations resulting from the specified case studies being influenced by ANT than to the potentials of the general research method ANT implies. Latour (1999: 20) has emphasized that ANT is a method, a way for social scientists to access sites, a way to travel from one spot to the next. This book utilizes the ANT as a methodological heuristic tool which has shown in concrete research practice how to treat humans and non-humans and micro- and macro-actors simultaneously on the same footing. Latour (1999: 22) has recently characterized ANT as a theory of the space or fluids circulating in a non-modern situation. ANT analyses technology always in relation to other elements of the network. The spatial view of technology which follows from such methodology has influenced our understanding of social spaces in working life, where technology is essentially a part of the construction of spaces.

Communities of practice

Changes in employment relationships and organizational forms influence the social setting of work in many ways. The increasing number of temporary work contracts, telework, the organization of work as projects or working in networks with partners in distant workplaces – each breaks the customary boundaries of workplace communities. Questions of daily forms of social interaction and new forms of communication have been acute concerns in the context of new spaces created through information and communication technologies. Revival of the interest in "communities" has brought under review both questions of actual working in changing social situations and broader questions of the basis of identification when customary communities based on workplace and occupational groups are replaced by more diffuse communities.

Rheingold (1995: 6) asserts that people around the world have a hunger for communities, as more and more informal public spaces disappear from our real life.

He talks about virtual communities, which according to him are "social aggrega-
tions that emerge from the Net when enough people carry on those public
discussions long enough, with sufficient personal feelings, to form webs of personal
relationships in cyberspace" (1995: 5) The word "community" is in this definition in
a very different context than, for example, in the classical texts of sociology, such as
Tönnies' (1967) "Gemeinschaft und Gesellschaft" and writings which have followed
from his differentiation between the notions of community and society/association.
Still, the underlying tone, both in the classical writings and in the present-day
discussion on virtual communities, is the longing for social settings which bind
people to meaningful social interaction and which are constitutive of their identities.

Jones (1995) has analysed new forms of community brought about by
computer-mediated communication (CMC). He links such new forms to the
potential of CMC for the production of social space. For him "CMC is, in
essence, socially produced space" (1995: 17). To clarify this he defines (1995: 16)
CMC as "not only structures of social relations, it is the space within which the
relations occur and the tool that individuals use to enter that space". Jones
remarks that in the history of community studies, space has been understood less
as socially produced and more as that which produces social relations. As an
example of this he refers to Stacey's (1974) identifications of the threads running
through definitions of community in sociological studies. These include territory,
social system and sense of belonging, territory being a boundary within which a
community maintains the other two elements. Jones (1995: 23) is asking for a
concomitant conceptualization of space and the social, an inquiry into the
connections between social relations, spatial practice, values and beliefs.

Jones remarks that the spur for development is connection, linkage. But this
connection is different from that interaction and togetherness which is predi-
cated on the nostalgic ideals of communities. Jones leaves open the question of
whether or not the communities formed by way of CMC should be understood
as communities. At least one can say that the concept of community is gaining
new content along with the development of new forms of connectedness and
mutuality. In common usage, the meaning of the word "community" has shifted
to a more symbolic direction, to refer to systems of social interaction, which to
an increasing extent are based on a symbolic sense of belonging, as for example
Lehtonen (1990: 242–3) points out.

Communities in working life do not change simply because of technology but
rather because of new work demands and related organizational changes, technology being one medium in the organizational arrangements. In organization
studies the social setting of work is seen to be crucial for achieving the objectives
of the organization, and since the Hawthorne studies (Roethlisberger and
Dickson 1949) the social setting has been viewed with the understanding that
formal and informal organizations interact. Increased co-operation requirements
crossing the customary functional fields and hierarchical levels within the organizations and moving boundaries between organizations, as well as increased
emphasis on knowledge as the key resource, have brought questions of community to the fore in a new light.

Nonaka and Takeuchi (1995) and Nonaka and Konno (1998) have dealt with the problem of the social setting of interaction in the context of knowledge management. They use a Japanese word, *ba*, to refer to the field of social interaction or, as they say, to the "shared space for emerging relationships" (1998: 40). According to them, the concept of *ba* unifies the physical space, the virtual space and the mental space. In their usage *ba* relates especially to knowledge creation, providing a platform for advancing individual and/or collective knowledge. The writers have described knowledge creation as a spiralling process of interactions between explicit and tacit knowledge. They differentiate four types of *ba*, which correspond to the four steps in the knowledge conversion process. Socialization involves the sharing of tacit knowledge between individuals. Originating *ba* offers a platform for this stage. In "originating *ba*" the writers refer to the world where individuals share feelings, emotions, experiences and mental models. They consider this the primary *ba* from which the knowledge creation process begins. Externalization, which requires the expression of tacit knowledge and its translation into comprehensible forms, is the next stage in the knowledge conversion process. It is supported by interacting *ba*, which is more consciously constructed as compared to the originating *ba*. For example, project teams composed of the right mix of specific knowledge and cross-functional teams provide conditions for such dialogue, which helps to make tacit knowledge explicit. Combination involves the conversion of explicit knowledge into more complex sets of explicit knowledge. Cyber *ba* is a place of interaction in a virtual world and corresponds to the combination stage. The writers think that the combining of new explicit knowledge with existing information and knowledge is most efficiently supported in collaborative environments utilizing information technology. Internalization produces operational knowledge by converting explicit knowledge into the organization's tacit knowledge. Exercising *ba* supports the internalization process by providing real-life or simulated situations for the use of explicit knowledge.

Tuomi (1999: 331) criticizes the model for ignoring the question, "What is the motive for knowledge creation?" Further, he remarks that Nonaka and his collaborators have not discussed which processes create the shared cognitive worlds, or *ba*s. He also sees problems in applying the model in situations where the spiral hits the boundaries of meaning creation space (1999: 328).

Wenger and Snyder (2000) approach the last problem in specified contexts with the concept of communities of practice. They overcome some of the problems in Nonaka and his collaborators' model by setting knowledge more clearly in relation to social practice. Wenger and Snyder ask how people share knowledge within and across organizations, and consider how it would be possible to foster new ways of knowledge sharing and learning. Their answer is through communities of practice, which are "groups of people informally bound together by shared expertise and passion for a joint enterprise" (p. 139). The concept of communities of practice has been developed by Lave and Wenger (1991: 98), who define it in the following way:

A community of practice is a set of relations among persons, activity, and world, over time and in relation with other tangential and overlapping communities of practice. A community of practice is an intrinsic condition for the existence of knowledge, not least because it provides the interpretive support for making sense of its heritage.

Originally Lave and Wenger used the concept to study situated learning in different kinds of communities. That approach took communities as stable entities, the interest being in socialization into a given cultural practice. Since Wenger (1998) and his colleagues shifted their interest also to the area of knowledge creation, the concept of communities of practice has taken on a more fluid and open content. Communities of practice are, according to them, fundamentally informal and self-organizing. What differentiates them from other forms of informal networks is that they are entities especially for developing members' capabilities and for maintaining and producing social stocks of knowledge.

Tuomi has suggested a broader definition of communities of practice. He has been concerned with how to combine the processes of knowledge generation at the community level and the accountability that is needed for the organizational level distribution of work and responsibilities. This is a management dilemma that has also troubled Brown and Duguid (2000). In practical organizational settings, teams approximate to communities. In contrast to communities of practice as defined by Wenger and his colleagues, teams are usually set up by a decision-maker and they have specified goals. Tuomi (1999: 398–400) suggests extending the concept of a team to include a periphery that is not responsible for the team goals, and the concept of community of practice so that teams can be community members, calling the new collectivity an "organizational community".

For the purposes of this book, Tuomi's suggestion makes sense. The book addresses questions of competence and identity. According to the understanding adopted here, these are questions which need to be dealt with in relation to communities. In organizational practices, both formal and informal aspects are present in communities. Nonaka and his collaborators' as well as Wenger and his collaborators' work reminds us of the fluid and dynamic nature of communities which exist in present-day organizations, and in spaces linking organizations. These communities provide a different basis for identification than more stable communities.

An agency perspective

The ANT approach looks at the question of agency in terms of results and effects and uses a broad definition: both humans and non-humans shape our social life and can from that perspective be considered as actors. When the focus is especially on human agency, as in this book, we need to elaborate the concept further. Giddens summarizes the core meaning of agency with two words: "power" and "choice". These are a good starting point to empirically approach constraints on and possibilities for action provided by working life. For Giddens

(1984: 9, 14–15), agency concerns events which an individual perpetrates, first in the sense that s/he could have acted differently, and second in that whatever happened would not have happened if that individual had not intervened. The capability of the individual to "make a difference" in a pre-existing state of affairs or course of action means that s/he exercises some sort of power. Giddens speaks in this definition about individuals, but the properties, power and choice, also apply to collective agency. Sociocultural and technological constraints and possibilities for action influence personal agency through self-processes. Central questions related to these processes are whether people believe in their capabilities to organize and execute the courses of action required to manage prospective situations (Bandura 1995) and whether people experience themselves as initiators of their own behaviour, being able to select desired outcomes and choose how to achieve them (Deci and Ryan 1987: 1025).

In this book, possibilities, choices and constraints on action are looked at through the lived world of the actors who in the book are blue- and white-collar workers, experts and managers from different sectors and organizational settings. The methods used in the cases are interviews, action research and organizational ethnographies.

Part I introduces and provides examples of the creation of spaces. The chapters discuss the role of ICT experts as creators of spaces, possibilities for supporting local agency based on regional information society strategy and problems of Web-mediated co-operation.

Tarja Tiainen's chapter analyses how ICT experts describe the process of creating technology. The subject of the study is ICT experts' views, since their work is to create information technology tools and services. Her chapter focuses on actors in technology creation: what kinds of human and non-human actors the ICT experts deal with and what kind of action space they ascribe to different actors. It is based on male and female ICT experts' interviews, in which they shared their visions of the future. The interviewees were specialists, consultants and managers in different areas of ICT. The interviewees' visions include several views on actors and action space. Contrasting views are those that follow the idea of technological determinism, in which people's action space is very limited, and those that see the shaping of technology as a complicated process in which human actors, organizations and technology are involved. The analysis of these alternative views of the creating of technology describes the shaping of human agency from ICT experts' point of view.

Marja Vehviläinen's chapter explores local agency in the information society. It starts from a concrete case of information society development in a peripheral region of Finland and the European Union, North Karelia. It looks at regional information society strategies, development projects run by the regional authorities, and Nordic welfare state and other institutional agents, especially those related to women's information technologies, as well as the lived experience of the inhabitants. The North Karelian information society strategy aims to create space for the citizens in the information society. Many of the development projects, funded partly by the EU and partly by national funds, aim to build facilities for the inhabitants to be active agents of the society, in their local and situated settings. The chapter discusses the construction of the social space of agency, with a special focus on gendering processes and the women inhabitants of the North Karelian information society. It examines the practices of the active agents, authorities, welfare state, institutions and inhabitants, and the network of North Karelian information society development. Further, it looks for the social and gendering orders of local agency in the information society on and across the borders of paid and voluntary work.

In Riitta Kuusinen's chapter, the stress is on an action research project related to a European programme on anticipating the form of future working life. The programme initiated a large number of projects with the expectation of generating co-operation between them. In addition, the projects had their own aims in their own organizations. The researcher was searching for a new action model for communication in virtual space within the framework of the EU programme and the guidelines of the ministries responsible for the organization of the programme. The process shows that the resultant virtual space was not enough for the creation of joint knowledge among the projects. The creation together of new practices for handling information and producing shared knowledge are difficult processes that need time and patience. What is needed is social support for the members of the group. This is possible only through the appropriate organization of social action and joint action learning.

Part II examines how people meet demands in new or transforming spaces. The empirical cases cover high-tech knowledge work, office work, service work, manufacturing plants facing local effects of globalization, and telework.

Sirpa Kolehmainen's chapter concentrates on work in newly developed, rapidly growing technology-based business services supporting the performance and success of the telecommunication industry. The focus is on organization and patterns of work at the workplace level in firms supplying services that are knowledge-based and produced by professional experts. High-tech professional knowledge work is characterized by complex problem-solving, based on both general and specific competence as well as analytical and social skills. While service products are usually tailored to the needs of clients, the service process calls for intensive interaction among the producer, possible co-producers and client. Thus, knowledge-intensive business service firms can often be called project organizations. This means that the work is arranged on a team basis along with the number of service projects. Individual workers usually participate in several projects at the same time, but their positions and tasks within each project team might vary. Project-based work organizations challenge the traditional forms of work as well as traditional conceptions of working communities. The chapter discusses the mobile boundaries of horizontal and vertical divisions of labour and the construction of the social community within knowledge-intensive project organizations. It also examines project-based work organization from the standpoint of continuous maintenance and the development of the knowledge and competences of an individual worker, as well as insecurity of employment and working conditions.

Pernilla Gripenberg's chapter on the virtualizing of the office takes a micro-level perspective on the impact of increased ICT use for people working with extensive use of office technology in office contexts. She identifies the web of forces and processes that sustain and drive increased ICT use in this setting. These include the importance of technology enthusiasts and a generally technology supportive culture, as well as the emotions that sustain and drive, or frustrate, the increased use of ICT in the office. She identifies some of the effects of the increased ICT use, raising questions of what media to use, when and how, what information to trust, and how to evaluate and store corporate information. Furthermore, changing patterns of interaction between organizational members and changing perceptions of time were identified as effects of increased ICT use. In the chapter, the term "virtualization" is used to describe the iterative, micro-level processes between ICT use, learning and institutionalization of ICT, so that it in turn becomes the organizational context of both the user and the use, again changing understandings of the ICT. In sum, the micro-level processes and drivers of increased ICT use can be understood as the result of processes of social interaction and learning between individuals.

Päivi Korvajärvi's chapter focuses on the creation of social communities of work that is based on the attraction and pleasure of ICT in female-dominated work. The analysis and interpretation are based on an ethnographic follow-up study conducted in a call centre. The case study shows that both management and employees stress the importance of the social community of work, although very differently. Management emphasized a pleasant atmosphere in the firm. When recruiting new employees, management evaluated how the candidate could "as a human being" cope with the social community of work. The employees' point of view was different from this. They stressed the significance of the social community of work as a supportive resource in concrete work. The employees also stressed the visions that the ICT opened up for them. In addition, they found that, because of the ICT, they were a part of the larger process of change in working life, which in itself is exciting. This gave them pleasure and the option to commit to their job and to the firm. Thus the ICT represented a kind of symbolic umbrella for the creation of the social community of work. Theoretically the chapter discusses how the emotional work required by the management and the commitment of the employees results in a social community of work. This acts as a resource for both managers and employees in different ways. For managers it is a resource in doing emotional work as a part of management, and for employees it is a resource that constitutes pleasurable involvement in their work. More broadly, the aim is to develop a dialogue between the concepts of "emotional work" and "compliance" in relation to the use of advanced ICT in female-dominated work.

Riitta Lavikka's chapter addresses female and male employees' internal mental processes in constructing the self in the emerging knowledge work of traditional manufacturing industry. The framework of the story comprises the structural changes at work created by the economic imperative of an information economy stressing individuality. Knowledge intensification of work, with the demand for individual responsibility and commitment, transforms employees' ways of being at work. If the work becomes a means of fulfilling an individual's own personal needs for development, its positive meaning for employees grows. It might also intensify the work to the point of exhaustion. The chapter aims to show, in the light of empirical research, what effects this two-way pressure creates in employees' orientations in work and life. The analysis is based on ethnographic interviews of female and male employees representing different hierarchical groups (shop floor and office workers, middle management, management) and observations of work gathered in eight manufacturing plants in different fields of production. The data reveal what kinds of subjective orientations lie behind the long-hours work culture. Knowledge intensification at work is understood here as a broad and general trend, describing changes in different aspects of working life connected to the growing importance of knowledge in the information economy. The importance of knowledge increases in all kinds of work, although the knowledge intensity of the jobs varies. The demand for individual responsibility and commitment, as well as the growth of social interaction at work, intertwines with knowledge intensity. The analysis also questions the traditional understanding of the central collective concepts of sociology of work, which tend to become problematic in analysing the work orientation of employees in the information society. A new approach is needed that includes employees' processes in creating a new sense of self and new identities that are connected to changes at work, new know-how and the co-operative dynamic of production.

The central concern of Riikka Kivimäki's chapter is on changing concrete practices in everyday life in the information society. The chapter analyses the meaning of working time and workplace in relation to the totality of life with interview data from teleworkers, mainly from new media occupations. Some of the interviewees work as teleworkers only part time, some full time. Within the totality of life, the spheres of working life and non-working life are not separate but linked together with many continuously reshaping ties. In the Fordist work organization, working time and working place have not taken many alternative forms. Shifting boundaries in working time and place mean changes in the interconnection and relationship of the pieces of the everyday life puzzle in the information society. On the one hand, this gives people new options that vary a lot according to life situations; on the other hand, the disappearance of boundaries brings its own problems. The relationship between family community and work community is in transition. Some new forms of communities can emerge in virtual networks. There can also be further challenges and possibilities in terms of new roles of women and men in families, work and networks.

The concluding discussion draws together experiences from the cases and reflects on them against the backdrop of discussions around "information society". The key aim of the book is the analysis of equal opportunities, agency and policy within an/the information society, and thereby the opening up of critical perspectives in mainstream public discussions.

Part II
The creation of spaces

2 Bounded or empowered by technology?

Information system specialists' views on people's freedom within technology

Tarja Tiainen

Introduction

Nowadays information and communication technology (ICT) is used by almost everyone in Finland. In this chapter, I focus on the creation and shaping of technology, and especially on the role various people play in the negotiation of its development.

Different views on technology creation suggest different roles for people. Technological determinism portrays technology as an exogenous and autonomous development, which constrains and determines social and economic organizations and relationships (Grint and Woolgar 1997: 11), so the individual's power is very limited. A second view of technology creation is that of an "organizational imperative", as Markus and Robey (1988) call it. This view assumes that systems designers can manage the impacts of information systems (IS). A third view is the social shaping of technology, which emphasizes the social, political and economic construction of technology. In addition, it shows that technology is constructed not only during its development, but also during its marketing and use (Vehviläinen 2000). This last view includes the idea that ICT contributes to employee empowerment as a powerful tool that can bring relevant information to the person on the front line. This view is used, for example, in envisioning telework and marketing ICT for it by saying that people can decide where they carry out their work (i.e. in an office or at home). The term "empowerment" also has another meaning, which emphasizes the rights and abilities of people to participate as equals in decisions about affairs that affect them (Clement 1994). ICT is presented as promoting this, too, as in the discussion of teledemocracy.

ICT and IS are generally developed in multiprofessional projects including IS specialists, which is a title that includes several groups of people. Dahlbom and Mathiassen (1997) outline the following categories of IS specialists: (1) programmers in the software industry; (2) support personnel in IS functions; and (3) management and trade union consultants who use computer technology to change things. The other groups of professionals in IS development projects come from the organization that needs a prospective IS. The participants are managers and users of the future system.

IS specialists characteristically dominate over the other occupational groups in IS development. IS specialists and users are talked about as opposite groups. However, the boundary between IS specialists and IS users is far from given or fixed, is fluid, negotiated, constructed, managed and, indeed, configured (Mackay *et al.* 2000: 739). My analysis of IS specialists' talk and users' action space in it describes one aspect of this negotiation. However, the negotiation situation is not equal, since the historical situation in IS projects gives a dominating position to IS specialists. In addition, the technology-oriented discussion about the information society underlines the importance of new ICT and it gives to IS specialists power to define the action space of users.

This chapter considers the negotiation of future technology from the IS specialists' point of view. I analyse how they talk about creating and shaping technology, and what kinds of roles people play in this context. One extreme case is that people do not actually have any action freedom in shaping technology, but they are bounded by it; the technology determinates the way of action. In the opposite case, people shape technology for their own purposes: people are empowered by technology. The aim of this chapter is to find out whether IS specialists expand or limit the possibilities for people to participate in technology development.

First, I briefly describe the central concepts I use in this chapter, outlining the characteristics of IS specialists and the context of negotiation between IS specialists and users. Second, I describe the research method I used. This is an interpretive study based on empirical material consisting of interviews with IS specialists. Third, I present the results of the analysis, which shows the general way in which the interviewees talk. After presenting these common ways of talking, I discuss all of them and highlight some mechanisms which limit people's freedom to act.

Central concepts

The central concepts of this study act as the lenses through which I looked at the empirical material. The first set of concepts describes the high value IS specialists place on the concepts of professionalism and expertise. The second set of concepts outlines different views of people focused on in this chapter. Finally, I deal with the information society, the context in which the negotiation of future ICT happens.

Professionalism and expertise

The projects in which ICT and IS are taken into use in an organizational situation are concrete, practical problem-solving projects. As professionalism involves specialization in a narrow field, multiprofessional co-operation is common. However, co-operation in multiprofessional groups is more problematic than acting in uniprofessional groups. In her study of multiprofessional groups in the medical field, Launis (1997) noticed that medical professionals overrespect others' specialities; there exist strict borders between specialities, and professionals do not cross these borders. This leads to the situation where professionals just talk about their own speciality and there is almost no dialogue between professionals on different specialities. The aim of multiprofessional co-operation is to understand the whole situation by getting information from several perspectives, but it is not reached. Launis (1997) suggests, as a solution, concrete co-operation in which real practical problems are solved.

IS development projects attempt to solve real practical problems by co-operation among professionals from several fields. At least two kinds of specialists participate in IS projects: (1) the members of the organization, those who need the new IS or ICT and who will benefit from it; and (2) IS specialists who are technology-oriented people designing and mediating IS and ICT. However, these projects are not simply unproblematic, equal co-operation between specialists from multiple fields, but typically, in them, technological knowledge dominates over work knowledge (Greenbaum and Kyng 1991). This problem has not been resolved in the mainstream IS development projects, although several research programmes have developed alternative practices in IS development. One of the alternative approaches is participatory design, which emphasizes the importance of the work context (e.g. Ehn 1988; Greenbaum and Kyng 1991; Bjerknes and Bratteteig 1995) and another is feminist system design, in which the one-sided technical expertise is problematized (e.g. Mörtberg 1997; Vehviläinen 1997; Karasti 2001).

In multiprofessional co-operation, both medical professionals and IS specialists are seldom able to reach an understanding of the whole situation. In these two cases the reason is different: medical professionals overrespect other professionals and IS specialists overlook others. In analysing the cases within the framework of professionalism, the professional status of medical professionals and IS specialists is different. Medical professionals are so-called traditional professionals; IS specialists can be called new professionals. The features of traditional professionalism include the fact that the professional knowledge is based on science, and the learning of professional knowledge in specialized (academic) education is a prerequisite to getting into the profession; furthermore, professionals belong to their professional community (Konttinen 1997). Although traditional professionals have an education in a specialized field, they, especially medical doctors, have been treated as "generalized wise men" (MacKinlay 1973).

IS specialists have such a visible and powerful position in the public discussion of the information society that it seems they are seen as "generalized wise men", as doctors used to be. Nevertheless, the case of IS specialists differs from traditional professionals in many ways. For a start, IS specialists' educational background varies; it can be information and computer sciences, or equally mathematics, engineering or business administration, for example, and some IS specialists do not have any formal education. Thus education does not give an entrée to the IS specialist community. The community is important for novice specialists to learn norms and for experienced specialists to update their knowledge in the rapidly changing ICT field. Besides novice and experienced IS specialists, the peer community plays an important role for those whose hobby is connected to computers, such as for hackers (Håpnes and Sørensen 1995). The case of hackers and computers as a hobby makes visible one difference between IS specialists and traditional professionals: the boundaries of the category of IS specialists are ambiguous or flexible.

The difference between traditional professionals and IS specialists can be seen from another perspective. Traditional professionals stay in their own domain, and customers leave their domain and reach into the professional domain, as is the case with patients and medical doctors; patients come to the doctors, who solve patients' problems from the medical point of view. The case of IS specialists is different. They leave their own domain and go to users' domains, and in the solving of users' problems, IS specialists are also expected to consider the user point of view. However, IS specialists have their own domain and their practices differ from users' practices. For example, IS specialists have their own jargon and they value innovation and technology itself highly, whereas users focus on the benefits that new technology offers for their tasks or to the organization (Orlikowski and Gash 1994; Davidson *et al.* 2001).

Despite the fact that IS specialists differ somehow from the picture of traditional professionals, however, they like to be valued as professionals. IS specialists commonly call themselves IS or computing professionals, as in the scientific journal *Communications of the ACM*, for example (e.g. Dahlbom and Mathiassen 1997; Denning 2001). Furthermore, like many professionals, IS specialists have adopted codes of ethics. Historically, professional associations have viewed codes of ethics as mechanisms to establish their status as a profession (Anderson *et al.* 1993).

Instead of talking about professionalism among IS specialists, expertise is another framework through which they can be viewed. Experts differ from non-experts in having a very deep understanding of the particular systems they are working with. The experts address problems, whereas the experienced non-expert carries out practised routines. Often these routines are carried out very well and are effective in a majority of cases. It is only when the routines fail that the difference between experts and non-experts becomes manifest (Bereiter and Scardamalia 1993: 9–16). In this framework, IS specialists are people whose understanding of IS is deep and who can manage IS in unexpected situations.

There are two paths on which one's expert career can develop. First, applying expert skills to broader social needs. As expert knowledge increases, so does the power to do good and to do harm; this includes the belief that knowledge is power. Instead of increasing specialization, some experts start broadening the use of their talent into other areas of value to the community. The second path is giving expertise away. This is the polar opposite of what professions are often accused of. Instead of hoarding specialized knowledge and making the public more dependent on it, expert knowledge is shared in order to make the public less dependent on them. The ideal would be for professions to do away with themselves by disseminating their knowledge or the fruits of their knowledge until there was no longer a need for their specialized expertise. This path represents progressive advancement in solving the problem of a field where progress means going beyond the individual delivery of professional services, and trying to do something about the conditions that give rise to a need for those services in the first place (Bereiter and Scardamalia 1993: 23–4). In the case of IS specialists, the first path towards an expert career means that IS specialists acquire knowledge in addition to IS and the technology domain, as well as about users' domains (organizational issues, for example). The second path means that IS specialists let users come to the IS domain and help users to manage the issues that require IS knowledge. The first path can be seen as users' boundedness, the second path as users' empowerment.

In this section, I have characterized IS specialists by using the concepts of professionalism and expertise. Both these concepts include some features of an IS specialist, thus there are some points in which IS specialists do not fit the ideal form of the concepts.

Humans, people and users

The terms that are used to refer to human beings vary according to discipline or context. The disciplines in which this study belongs are IS studies and social studies of technology. The terms I use to refer to human beings are "human", "people" and "individual", and I use two further categories, "IS specialists" and "users". Now I discuss their meanings, boundaries and the underlying assumptions they include.

First, I take the terms "human" and "non-human". According to Grint and Woolgar (1997), common to all different definitions of technology is the attempt to distinguish between human and non-human elements. "Non-human" tends to be associated with the material, intrinsic, technical content, whereas "human" tends to connote the (merely) circumstantial context (social factors). However, despite the implicit tendency of many authors to fashion their discussions of technology in terms of this dichotomy, it is difficult to sustain the boundary between human and non-human. Humans do not act without some form of "artificial" (i.e. humanly constructed) construction (clothes, tools, machines, etc.) and non-humans of this "artificial" form do not act in the absence of humans (Grint and Woolgar 1997: 9–10).

In the IS field, the category "humans" is divided into two separate groups: IS specialists and users. IS specialists are those who develop and create ICT and IS, while users use the finished technology without shaping it. The dichotomy between active technology shapers and passive technology users is the dominant image in the IS field (e.g. Ives *et al.* 1980; Avgerou 2000). However, the boundary between the two groups is fuzzy. IS specialists do not just develop ICT and IS, they also use ICT and IS themselves. They write reports and memos, seek information from the Internet, send e-mails, and use computer-aided tools for IS development; nevertheless, they are not called users. Furthermore, there are users who develop their own systems. This issue is called end-user computing, which is defined (Brancheau and Brown 1993: 439) as: "the adoption and use of information technology by personnel outside the information systems department to develop software applications in support of organizational tasks", emphasizing the word "develop".

The above division into IS specialists and users is based on the separation of designing (or shaping) and using ICT. The boundary between them is not sharp, as the above example of end-user computing showed. In addition, various social constructivist approaches to technology argue that technology is also shaped during its use (e.g. Grint and Woolgar 1997; Orlikowski 2000). Furthermore, some alternative schools of IS development challenge the dichotomy between use and development, for example, in participatory design (Ehn 1988; Bjerknes and Bratteteig 1995) and feminist IS development (e.g. Mörtberg 1997; Vehviläinen 1997; Karasti 2001).

In the IS field, it is normal to use the term "users" to refer to people who use computers. Greenbaum and Kyng (1991: 3) criticize the term: "The term lumps all kinds of workplace activity together implicitly putting the computer in focus and treating people as a blurred background." Besides putting technology in focus, the term "user" hides the expertise that others than IS specialists have – the others might be financial managers and accountants, for example, or doctors and trained nurses in the case of medical IS. Sometimes the users are not experts, as in the case of the electronic business, when IS is made for everyone to buy products via the Internet. In such cases, IS specialists (and IS researchers) call those people "users", but someone else may call them "laypersons", "consumers" or "citizens".

Besides the terms presented above, I also use "people" and "individual". These are not contradictory, as the earlier pairs of terms were. "People" and "individual" refer to quite similar kinds of persons. The term "people" refers to anybody. It is used in cases in which it is no matter who the person actually is. It means a kind of stereotypical person. Instead of using "people", I use "individual" when I want to underline that the person performs alone, independently of others.

The information society

Discussion of the information society is the context in which the negotiation of technology development happens. The information society discussion during the last decade underlines the importance of ICT. This discussion strengthens IS specialists' position in defining users' action space within technology, so here I outline its main characteristics.

The discussion about the information society started over thirty years ago (e.g. Drucker's book *The Age of Discontinuity* (1969)), but the discussion became more common, notable and technology-centred in the 1990s. A remarkable milestone was reached in 1993 when President Bill Clinton and Vice-President Al Gore linked the US's destiny with the creation of the National Information Infrastructure (NII): that is, a vision of universal access to seamless computer networks and unlimited amounts of information. European Union and European countries followed this technology-centred line in the creating of their strategies towards the information society (Karvonen 2000b), as did cities in Finland (Eriksson 1999; Lehtimäki 1999).

In Finland, public discussion about the information society is technology-driven. The attitude towards technology is optimistic: there exists technological determinism (Aro 2000), and Finnish technology creators are described as heroes; examples include Jorma Ollila, Nokia CEO, and Linus Torvalds, the 'father' of the operating system Linux (e.g. Lyytinen and Goodman 1999). New ICT is put at the centre in the discussion of the information society; however, this view is challenged by the social shaping of technology approach, which sees individual people's actions as important (MacKenzie and Wajcman 1999). Following this theoretical viewpoint, Eriksson and Vehviläinen (1999) and Uotinen *et al.* (2001) have challenged the Finnish technology-centred discussion about the information society, focusing on the citizen point of view on ICT using and shaping.

Despite some marginal human-centred voices in the discussion of the information society, the main discussion is technology-centred. This kind of discussion strengthens the position of IS specialists, since they are specialized in technology and they mediate it to others. In this frame they are the core actors, not only in technology shaping but also in the shaping of the information society.

Methods

IS specialists are very visible in the discussion of future technology. As they dominate others in IS projects, I have examined whether a similar pattern is to be seen in the way they talk about technology shaping. I analyse how IS specialists' talk about technology creating and shaping, and how people are related to these processes. My aim is to discover the ways of talking that expand and constrain people's freedom within ICT. The empirical material that I use includes interviews of IS specialists. The analysis makes visible something of the negotiation of participants' roles in IS development projects, or in ICT and IS shaping.

I interviewed twenty-four IS specialists who are not programmers but have upper-level posts. Among the interviewees were managers and decision-makers within ICT, and mediators of ICT and IS. The interviewees live in Finnish society, which is aiming at becoming one of the first information societies. Many of the interviewees have contacts abroad; besides present business contacts, some of them have lived – studied or worked – in other countries.

The interviewees were chosen from different types of organizations and use ICT and IS in various ways. One criterion in choosing them was that their work should be future-oriented or that they should be developing some novelties in a technical or social sense. One part of their work is discussion. They present their ideas within or outside their companies. They also participate in several negotiations with customers, colleagues or policy-makers. As well as this, many are visible in the public discussions of future ICT. In the interview situation I encountered their skilfulness in discussion.

In the interviews I asked the interviewees about their visions of the future, using open questions. The themes that I asked about included future changes in technology, people's lives, organizations, society, and the global situation. The interview situation was a peer discussion, since my background is in computer and information sciences and I have working experience in practical IS work in a software house, though I have not done the same work as the interviewees. Being a peer of the interviewees facilitated the interviews, since I share the interviewees' working culture and may understand the language used there. Instead of controlling the progress of the interviews, my role was as an interlocutor, who accepted their views and opinions but whose aim was to get them to speak more and expand their statements.

In the interviews, most of the interviewees behaved as though they were sure that their ideas were excellent and that they knew best. This feeling was especially strong with technical issues. When we reached to the social issues, some of the interviewees said that they were not experts in that area; nevertheless, they described their ideas in detail. I found that among the interviewees there were three exceptions to this. Two middle-aged women did not present themselves as technical experts. However, they assumed that an IS specialist should be such, and they asked for my acceptance of their visions. The third exception was a middle-aged man. Instead of being omniscient, he was pondering and looking for alternatives.

I analysed the interviewees' visions by using discourse analysis. I view discourse as an expression of cultural issues and have interpreted the material as quite local-situated. This means being relatively sensitive to language use in the context, but also being interested in finding broader patterns and going beyond the details of the text and generalizing to similar local contexts (Alvesson and Karreman 2000). There are two contexts in which the interviewees belong: first, the IS specialists' community, and second, the (Finnish) information society.

The cultural aspect of narration can be interrogated with critical discourse analysis. It places a lot of emphasis on the implied messages that underlie communication. It is assumed that ideological moulding takes place, not just through explicit information but also through the implied propositions that are brought to bear in trying to make sense of statements. Whatever is endowed with such self-evident truth that it does not need to be said aloud must be assumed to be correct. There are four degrees of presence and absence: (1) foreground information refers to those ideas that are present and emphasized; (2) background information refers to ideas that are explicitly stated but de-emphasized; (3) presupposed information is present at the level of implied or suggested meaning. Finally, one needs to consider (4) absent information, ideas and perspectives that are relevant to a topic but neither stated nor implied in a text (Riggins 1997). The last two degrees require the awareness of the context of the interviewees. In the analysis I have tried to be sensitive to all of these levels.

People in relation to technology

The shaping of technology for the future happens in negotiations among several groups of people. Discussion on the information society focuses on ICT and so strengthens IS specialists' power in shaping the future. As IS specialists are powerful in the negotiations on technological development, I analyse how they talk about other people, especially what kind of position other people have in shaping technology.

This analysis describes what kind of experts IS specialists are: whether they try to keep the IS field to themselves – as traditional professionals do – or let others come into the IS field, which is one possible developing path of expertise – "giving expertise away", as Bereiter and Scardamalia (1993: 23–4) call it. Furthermore, the analysis describes how IS specialists draw a distinction between themselves and others, especially users of ICT.

In this section I also present the common features found when analysing the interviewees' talk. A general way in which they talked was "planning talk", in which people are just the objects of planning. Planning talk is the dominant form of talking and all the interviewees used it at least some of the time. In addition, the interviewees' talk included some role for people, too. First, they told me that technology is used for remedying people's defects and improving people. In this case, people are also objects of action but now they are more visible than in planning talk. Second, some interviewees stated that people choose from ready technological artefacts which ones they will use.

This section is based on the interviews; I found the types presented below in the interview material. I give space for the interviewees' voices and use references only when making my interpretations of people's positions in the interviews more understandable. I discuss the findings of this analysis in the next two main sections.

Type 1a: planning talk about technology

The first impression from the interviews is that in them people are not viewed as active participants. The talk does include some references to people, but often their role is to be the objects of planning. In the creation of new technology, the schema is as follows: it happens that something – that is, better technology – comes along and has (positive) effects on people. This event is described as inevitable and also as desirable. No actors are mentioned who make this event happen.

Some examples of this kind of talk follow. The first quotation is taken from Kari's interview. Kari (a male consultant, aged about 50) states that, in future, life will be automated with technology and will be easier than life nowadays:

> Technology will be used for automation, for making people's everyday stuff easier, decreasing routine … A normal view seems to be that, if living, for example, is automated, then it becomes terribly complicated since there will

be terribly many buttons. So how can anyone figure out all the buttons? My opinion about automation is that it decreases the number of buttons, since not all are needed and no one needs to pay attention to the stuff that is automated. So things will become easier and they will need less knowledge when the buttons are taken away and the co-operation with the machine will decrease. That means things will, without our noticing, shift to the background and they will become automatic, so that people will have more time for co-operation together.

The above quotation concerns "a normal view". It states that the normal view includes a fear of new technology and that this view is wrong. This is a preface for showing the right view, according to which technology will make people's lives easier. It is convenient to present the future with such a combination – the incorrect normal view (held by users) and one's own right view (held by IS specialists). This way of talking strengthens IS specialists' power in technology-shaping and it decreases others' power.

Planning talk is common in the interview material. Another example is taken from Petri's interview, from a section where we are discussing future user interfaces. Petri (a male consultant in his mid-thirties) talks about his hopes for development:

This three-dimensionality and still greater realism, that is the imitating of everyday life, it might be a sensible direction for development. Like nowadays, it, the user interface, is just a plane, so it could be replaced by a three-dimensional space. But it's hard to say what kind it might be then. It needs some kind of a virtual helmet, since it'd be quite difficult from a tube [computer screen]. And, no doubt, in ten years it will be possible to get a functional one [a virtual helmet]. You'll have some kind of glasses, either just ordinary ones, inside which the picture is just reflected, or a bigger pot [headgear] that is put around the head, and there you'll see everything three-dimensionally and you can control it. This is a realistic direction and there might be a demand for it.

In this quotation, like the previous one, people are not mentioned directly, although they are somehow present (e.g. "there might be a demand for it"). The quotation states that future technology will become "realistic". Petri used the term "realistic" to mean that, in the future, acting in a virtual world will be like acting in the real world. The idea is that it is good if the using of a machine is like dealing with other people. Petri is not the only one of the interviewees who envisioned three–dimensional (3D) user interfaces. For example, Lauri (a male manager, about 50) described to me a similar vision. He argued that this kind of interface is good, since "if you can see a hundred objects with a two-dimensional [interface], then you can see ten thousand objects [with a 3D interface]". This statement includes the idea that increasing the quantity means increasing the quality.

I call this type of talking "planning talk". The meaning of planning talk can be clarified by drawing on Hearn and Roberts' (1976) use of the term in analysing social planning in health and welfare. They characterize planning as a political activity, operating to the advantage of existing power groups. Planning talk helps to rationalize plans and is preoccupied, above all, with references to objectives and priorities. One form of planning talk is client analysis of client groups, such as "the elderly" and "children". Such groupings may appear logical and sensible, but not all groups are identified and not all needs are presented as equally legitimate. They criticize such seemingly rational planning procedures, as by rigidly categorizing people and segmenting problems they tend to inhibit development of new thinking and creative policy.

I find that people are more visible in the context of health and welfare than in the case of future technology. In my interviews, the planning talk of technology focuses on technology and assumes that the development of technology is a good thing. IS specialists' talk characteristically includes the view that IS specialists are the right people to define what kind of technology is good for all people. Indeed, all interviewees used this kind of talk, but most of them also used other kinds of talk, at least sometimes.

Type 1b: technology for remedying defects and improving people

Another general assumption reflected in IS specialist talk is that technology is used to remedy some "defects" people have. This type of talking also fits into the frame of planning talk, in which people are objects, but now they are more visible than they were in the above type. Because of the compatibility between this type and the above type, I call these Type 1a and Type 1b.

Type 1b appears when the interviewees described ICT as perhaps being useful for aiding the handicapped and the elderly. They have something lacking or some defects that ICT could replace. An example can be taken from my interview with Liisa, in which we partly discussed the advantages and disadvantages of a future vision which includes a direct connection between a computer and the mind. Liisa (a female manager in the public sector, over 50) said that ICT could remedy blindness, but she also said that a healthy person must not be improved with ICT, since it might dispel people's humanity. This is what she said:

> A machine can replace some parts (of a person) ... Of course, as people have [presented] these forecasts of virtual reality, there is one remarkable application area of virtual reality, that sight [is returned], for example, or a tool for the blind, and this requires a connection [to the brain] so that the virtual reality is reflected in the blind person's so-called visual field. This, to my mind, is one of the issues in which virtual reality might be useful.

Usually the interviewees described the handicapped and the elderly as people who need aid from other people and whom ICT could help in the future. This kind of argument is connected to the discussion about the population of Finland getting older. Sometimes it is pondered who will take care of all the elderly. The interviewees contribute one opinion to this discussion.

Usually, interviewees who used Type 1b talk mentioned specified groups of people whose problems need to be remedied. Kari (the 50-year-old male consultant quoted above) is the only exception among the interviewees, since he said that everyone is defective and needs to be helped. He dealt with logical thinking; he said that everyone's logical thinking is defective and it needs to be improved. According to him, computers give an opportunity to educate people, since they can obtain more frequent and more exact feedback on their behaviour. Kari does not see any difference between disabled and "normal" people, since he views everyone as disabled.

Type 2: technology for individuals to choose

People are the target of planning in both the above types of talk (Types 1a and 1b). The interviews also include statements in which people are shown as active participants, with some freedom towards technology. The point in which this is shown explicitly is that ready technological artefacts are given for people to choose and to accept into use.

Whereas the above types of talk focus on the development process of technology, this type deals with the use of technology. Furthermore, in this type the users are not seen as passive receivers of technology, but are seen as participating in technology shaping. This type of talk – more than two above types – resembles the view underlying the social shaping of technology approach, which emphasizes the social, political and economic construction of technology. In addition, it shows that technology is not constructed only during its development, but also during its marketing and use (Vehviläinen 2000). Usually IS studies focus on the development process alone, as did most of the interviewed IS specialists, but Orlikowski (2000) is an exception among IS studies in focusing on the user's role in shaping technology.

As an example of this third type of talk, I take a quotation from Anne's interview. In it, Anne (a female manager in a multinational company, about 50) is describing telework and teleconferences, and the technology used in them. She explains that, nowadays, defective technology limits the use of telework; when it is sorted out, telework will be an available alternative more often.

Interviewer: Do you think that people will want to do it [to telework]?

Anne: Not everyone will. I think … that we'll give [an opportunity] to people. Those who want it, they'll get a chance. So, we'll create frameworks and opportunities. Different people like different things and ways of working. We'll give the opportunity to work in a more flexible and multiform way. I'm sure it [the use of telework] will increase and technology will help it.

Interviewer: And how is it with you? You are in the US every other month, so could it decrease the need for travelling there?

Anne: It is a question of type, who likes what. I'm such a people-oriented person that in my case it would not decrease it, since I find face-to-face contacts necessary. But, sure, it would decrease, since not everyone finds it so important or needs it in a similar way. On the other hand, it is very hard to travel, with the long flights, and some find it physically hard or do not feel good doing it. Such persons, who do not like travelling, it is easier for them to work effectively if they do not have to travel. Of course, it depends on the work. It is again an individual question.

Interviewer: You mean that there is more freedom?

Anne: Freedom to choose, or technology will give opportunities for it and probably it will be used.

The above quotation is not a typical example of interviewees' talk; rather, it is the opposite – only a minority of the interviewees had even one statement of this type. Furthermore, I found that, among the interviewees, Anne gave the largest action space to users in technology shaping. By users' action space in technology shaping, I mean the possibilities users have in affecting the development of technological artefacts and in deciding how those artefacts are used, especially in defining their own way of using the technology or in shaping technology for their own purposes.

In this way of talking, people are participating in technology development. Technological artefacts are first prepared (and it is not explicitly said that people participate in this process), then the finished artefacts are given to people who decide whether they select them and want to use them. When people can choose the artefacts they use, they can even reject some technological artefacts when they do not like them. Some interviewees describe such a situation: for example, some of them guess that maybe people do not like the vision telephone (i.e. telephone with a picture) and would never put it into use. Furthermore, people can reject a technological artefact if they think it is unethical, as someone thought that a direct contact from a computer to the human mind might be.

Ways to limit people's freedom to act

Above, I presented three ways in which the interviewed IS specialists talk about technology and people. In the talk that belongs to Types 1a and 1b, people are treated as objects of planning. In such cases people's freedom to act is very limited, since they are not thought to be active participants. Technology is changeable and actively developed, whereas people are passive and the changes in them come from the demands of technology. When the situation is analysed from those people's point of view, they are bounded by technology. The talk of Type 2 is almost the opposite. It includes a predisposition that people are active participants who can affect the development of technology so that it answers their needs and hopes. In such a case, they can use technology for empowerment.

Since the interviewees commonly used the talk of Types 1a and 1b while the talk of Type 2 was rare, I explored the mechanisms underlying how people's freedom to act within technology is limited in such talk. Above, I argued that the term "user", which is commonly used in the IS field, is problematic. When one group of negotiators is called "users" and the other "specialists" or "professionals", it includes a predisposition that the user group is marginal, less powerful. Furthermore, the talk about users' participation in development includes the predisposition that users' role is less important (Nurminen 1986, 1988). The other mechanism that is used is reference to "normal people", which is supposed to refer to everyone. This fits together with the use of planning talk. Now I turn to the views that were expressed on what was understood by "normal people".

It may be easier to come to the view of normal people from its opposite: what is an abnormal person? It is dealt with in one of the ways in which IS specialists talk: that is, describing technology to be used to remedy some defects that people have. Those people whose defects need to be remedied are abnormal and they become more normal when their defects are remedied. This kind of talking raises the question of the defectiveness and nondefectiveness of people, i.e. what kinds of people are regarded as normal or are accepted. Besides this type of talking, other types of talk also include some assumptions about people – the speaker has in mind a view of the person s/he talks about. The person in mind has characteristics (such as appearance, age and gender) and skills and attitudes (such as the ability and the will to use new technology). There are several human features, but I just deal here with three: health, use of technology, and gender. I chose these since they exist in interviews and they are discussed within ICT and IS more widely (e.g. Wajcman 1991; Kling 1996).

It is assumed that a normal person is healthy, which means that s/he does not have any disabilities. However, the concepts of health and disability are not unambiguous. A common way to understand disability is as the loss of a function in a part of body, and so rehabilitation and normalization are spoken of (Moser 2000). In Type 1b above (technology for remedying and improving people), I showed an example of blindness and the returning of sight, which was mentioned by Liisa. This example follows the idea of rehabilitation and normalization.

Moser (2000) challenges our ways of thinking about normal and abnormal, about ability and disability. Understanding disability as normalization marginalizes and excludes disabled people, but some other frames give more space to disabled people. One such thought-model is the cyborg, which is a hybrid of machine and organism, and certainly also of machine and human. Cyborgs symbolize that we have now become so thoroughly and radically merged and fused with the technology that it is impossible to say whether cyborgs are "really" human or "really" machine. This kind of thought-model views it as normal that people use ICT to improve their performance, and does not draw any distinction between disabled and "normal" people (Moser 2000).

Science fiction stories present one idea of cyborgs, and there the view is based on technological hype. In those stories, the cyborgs are created in science laboratories and the results are abnormal individuals with non-human skills or features, such as extraordinary power and tirelessness. An extreme example is the man in *Terminator*, he could turn into a form of fluid metal. Besides these imaginary views of cyborgs – hybrids of technology and human – cyborgs also exist in real life. Persons with heart pacemakers and artificial legs are examples. In such cases, the idea is that those persons' defects are remedied with technical artefacts, so they are abnormal people who are trying to be normalized – using Moser's terms. Besides those extraordinary examples, we can also put into the category of "cyborgs" persons with contact lenses or dental fillings. Those cases are so general that normally they are not thought of as remedying some defects.

Generally, when humans and technology are talked about, they are meant separately: not cyborgs, but humans using technical machines. In Finland, the dominant public discussion of the information society includes a clear picture of people as users of technology: it is assumed that all Finns are interested in technology, that all Finns use new ICT and want to use it, and that all Finns have skills for using it (Vehviläinen 2001b).

The view of people using new ICT was not very homogeneous in the interview material, but reflected the assumption that ideally people would be interested in technology, would use it and would have technological skills. The interviewees talked about different kinds of people who have different habits in using new ICT, but different habits were not presented as equally desirable. The interviewees described old women as a problem group, since they are not interested in using technology and they have no skills for it (Kuosa 2000). The other special group of people that some of the interviewees dealt with was the young. They were presented as heroes and their way of using technology was shown as the best one (see Kuosa 2001).

As mentioned above, some interviewees presented older women as a problem in ICT use. More generally, the interviewees described ICT as a male area (Kuosa 2000). In the interviews, as well as elsewhere, technological expertise and masculinity are connected on a symbolic level. Merete Lie (1995) describes this connection by understanding masculinity as an abstract frame of reference, a kind of standard one refers to in the articulation of one's own as well as other people's gender. The masculine symbols may not be reflections of the capacities of "real men" – which means that there exist some men who deviate from this picture – but more probably images of normative masculinity. Defining masculinity as a frame of reference and not equal to the behaviour of "real men" does not mean that it is less "real" or less important than actual behaviour. The level of conceptualization and symbolization of gender is just as real as what we observe men and women do (Lie 1995).

The consequence of the connection between masculinity and technology is that normal people are assumed to be male. This happens not only in the case of technical experts, but also when referring to users, which is shown in the case of presenting older women as problems. Furthermore, IS specialists are in the habit of thinking that the users of IS are as similar as the IS specialists are, especially in cases in which the users are unknown, as with packaged software. This view of users states that they are young, technology-interested men (Isomäki 1999; McDonough 1999; Rommes 2000).

Some of the users fit the above picture of normal people, but some do not. Those who do not fit to the picture of normal ones are "others". Riggins (1997) explains that others may be practically invisible, others can assimilate as a whole or in part, and others may be devalued but at the same time eroticized and envied. In the case of ICT, one group of others is women. Although in Finland women use ICT in their work, even a little more often than men do (Lehto and Sutela 1999), women are quite invisible in mastering technology (e.g. Vehviläinen 1997). In her analysis of a Finnish computing pioneer's autobiography, Vehviläinen (1999b) encountered a male pioneer drawing a picture of a world without women. Even nowadays, female IS specialists are especially rare in the Finnish news. The general discussion of the information society connotes assimilation for others. Means are sought to get the people in supposedly problematic groups – women and the elderly – to use new ICT, means to get them to act as technology-interested young men do.

Riggins (1997) says that it is hard to find any action space for someone who belongs to the category "others", since they are devalued, marginalized and silenced by dominant majorities. The views of normal people and others are conceptualized as a "production", which is never complete, always in process and always constituted within, not outside, representation. When the interviewees' talk is analysed in the context of production of normal people, it reaffirms existing views. The interviewees' talk in Types 1a (planning talk of technology) and 1b (technology for remedying defects and improving people) follows these lines, and such forms of talk limit people's freedom to act.

Also, the public discussion of the information society contributes to the production of the view of normal people. This discussion includes some proposals to modify the dominant view and give more space for others. One such attempt is a research project in North Karelia, which seeks space for alternative, local practices in using ICT (Vehviläinen 2001c, Ch. 3 this volume). As the starting point is a local situation and people's needs there, it is easier to remember that the groups of others – such as the disabled, women, the elderly – are not homogenous. An example of the versatility of those groups is the elderly. Although many elderly people have no experience of ICT, computing pioneers have retired and so they also belong to this group. The interviewees' talk of Type 2 (technology for individuals to choose) includes some space for divergence.

Concluding comments

In this chapter I have dealt with IS specialists' ways of talking about people in the context of technology creation and development. I presented three types of their talking about people: Type 1a, planning talk of technology; Type 1b, technology for remedying defects and improving people; and Type 2, technology for individuals to choose. These types include several of limiting people's freedom to act, which I discussed above. Now I return to those concepts presented at the beginning of this chapter, which outline the IS specialists' role and negotiations about future developments.

The first set of concepts presented includes professionalism and expertise. In the case of traditional professionalism, the boundaries between different professional domains are strict, when professionals respect others' domains and do not encroach upon them (Launis 1997). The case of IS specialists is different, since they work in multiprofessional groups. IS specialists do not stay in their own domain but go to the situation in which IS is needed, and so they get into the users' domain. In this situation, IS specialists used different ways of preventing others from taking over their professional area.

Besides professionalism, we can link the ways in which people's freedom is limited within technology to the framework of expertise. Bereiter and Scardamalia (1993: 23–4) describe two paths on which one's expert career can develop: (1) applying expert skills to broader social needs; and (2) giving expertise away. IS specialists cannot keep all the expertise of the IS field to themselves, whether they want to or not. During the last two decades computers have become easier to use and lots of software packages have appeared for anyone to buy (see e.g. Friedman and Cornford 1989: 231–43). As IS specialists use several ways of limiting people's freedom to act, these ways can be seen as IS specialists' reactions towards the changes and their attempt to control the loss of their expertise – or to prevent or delay this process.

The second set of concepts presented includes "humans", "people" and "users". In the IS field it is normal to use the term "user" to refer to people, regardless of whether or not this term involves a technology-centred focus (Greenbaum and Kyng 1991). It is an old idea that users' participation in IS development is important (e.g. Mumford and Henshall 1979), but one that has not yet been generally realized (e.g. Karasti 2001).

IS specialists know that users' participation is important and they explicitly affirm this when asked directly, but in indirect situations they might pass it over. Eteläpelto (1994) recognized this in the analysis of novice and expert IS specialists. Eteläpelto's study gives us hope: when someone acquires practical experience and moves from being a novice towards becoming an expert, the contextual situation and users are understood better. Unfortunately, my interview material does not support this view. The older and more experienced interviewees did not give noticeably larger action space for users than the younger ones.

Another explanation is to be found in Isomäki's (1999) study. She recognizes that IS specialists' view of human beings is very superficial. This finding fits together well with the results of my study: as IS specialists' view of people is superficial, it is not surprising that they do not see any need to expand people's freedom to act within technology. On the contrary, they are interested in technology and know it well, so it is easier to imagine its possibilities. My study, as well as Eteläpelto's, supports the idea that the real expertise in the IS field is demanding and requires multiplicity.

The last concept examined is that of the information society. It is the context in which IS specialists act and which reasserts IS specialists' power in defining development lines. Since IS specialists have a highly powerful position in the public discussion of the information society, it is important to study IS specialists' views on technology construction. While the results of this analysis present the idea that IS specialists limit other people's freedom to act in many ways, we can start to reflect on the existing ways in which the information society is discussed and developed. IS specialists have considerable power in this process.

This study has dealt with the question of IS specialists' power and the ways in which they keep power to themselves. The answer that this study gives is that users are put in an unsubstantial position of "others". How IS specialists understand differences between themselves and users in IS development needs further study. In the particular study discussed here, I have described IS specialists within the framework of professionalism and expertise. The same framework would appear to be useful for future studies of IS specialists' action and co-operation with users.

Acknowledgements

I am grateful for the helpful comments I received on earlier versions of this manuscript from Jeff Hearn, Hannakaisa Isomäki, Pertti Järvinen, Tuula Heiskanen, Karlheinz Kautz, John Law, Christina Mörtberg and Marja Vehviläinen. Also, I would like to thank Joan Löfgren for making my English more readable.

3 Gendered local agency

The development of regional information society

Marja Vehviläinen

People's relation to technically mediated society is often discussed in terms of access to technical artefacts, or maybe the distribution of artefacts in various groups of people, countries and continents. Furthermore, there is a large literature of the use of the Internet, mobile phones and other specific technical artefacts. Distinct from these approaches, this chapter explores the local agency of people in a technologically mediated society. Local agency is situated in people's everyday lives and practices, and is shaped within the social orders of the distribution and use of technologies. The chapter's aim is to give space to the situated practices and societal differences and, at the same time, articulate orders of technically mediated society from the starting point of these practices in a particular case: women's information technology groups established as a part of an information society development project being carried out in North Karelia on the eastern border of Finland.

People live and act locally, for example in North Karelia, but they do not make choices, on the use of technologies or the development of their activities independently from the rest of the social world. There are partially technologically and textually mediated social orders and relations present in their activity. Some of the relations are very local and are related to particular individuals and their subjectivities. For example, people live together with other people, maybe among family members, and they share television sets, computers and so on, as well as care for each other, or they study or work in small groups and share knowledge, practices and artefacts with the others. Some of the relations are broader in scope. People use the Internet on several continents and they get funding from national and transnational sources, which connects them to national, transnational – for example, EU – and global networks. Local agency is situated, and it intertwines with these orders (Haraway 1991). People act within the orders and relations and they also (re)define and (re)shape them – as they aim to do through the women's information technology groups discussed here. They also participate in the development of the technically mediated society. I examine these orders in order to understand the space for people's local and situated agency.

The development of new societal forms, often referred to as information society, puts agency and citizenship to the centre of both political and academic debates. Although the content of the concept of the information society varies from primarily technical development to cultural and global economic changes (e.g. Webster 1995), it nevertheless seems to involve changes in the agency in important domains of people's lives. Various writers (e.g. Compaine 2001; Loader 1998) have shown the threats of the digital divide: the agency is shaped according to the divide between North and South, well educated and poorly educated, urban and rural, young and old, as well as male and female gender. Some others (e.g. Escobar 1999) have set out to find spaces for agency for those on the "wrong" side of the divide. Both elements, the divide connected to the societal differences and the possibilities for agency, are present in my chapter, too.

Ruth Lister (1997) points out that agency and citizenship have two sides: the rights of citizenship that are universal to all, and the particular social practices that comprise the local and situated activity of people, organized in various social relations of gender, class, age and place. These practices shape citizenship and agency and intervene in people's possibilities of using their universal rights. In Finland, certain governmental programmes aim to build equal access to computers, the Internet and the information available through information and communication technologies (ICT) for all citizens. Equal access to information technology is a universal right for Finnish citizens. In the 1990s, libraries as well as various citizen and neighbourhood centres provided opportunities for computer use. Schools in most parts of Finland have good computing facilities, and Finland ranks highly within international measures of mobile phone, computer and Internet usage. Still, the actual agency and citizenship of people go beyond the right of equal access to the practices (and the orders and relations embedded in them), as well as to the subjectivity of people.

Although "equal access to computers" as intended in governmental policies is generally provided, the social orders present in people's agency make the question of equality complex (e.g. Henwood *et al.* 2000). Gender is an especially tricky relation within technologically mediated agency. Although both women and men use computers and the Internet fluently (Lehto and Sutela 1999), women tend to define themselves as non-experts (e.g. Tuuva 2000; Margolis and Fisher 2002). There seems to be a commonly shared knowledge in Western societies that it is certain men who have the core expertise in technology generally and in computing particularly. There is a nexus between (male) experts and technology, as Judy Wajcman (1991: 165) puts it; this undermines women's starting point within the technologically mediated activities.

This chapter acknowledges the rights of citizens and traces them from the policy documents of both national and regional authorities. The focus is, however, on the situated practices of people. I examine how local actors shape their gendered agency in their local setting, within the rights and social orders of the situated practices.

The study was carried out in the province of North Karelia, on the eastern border of Finland and thus of the European Union. The region has had high unemployment (at the time 20 per cent, in parts 25–30 per cent) and long distances which had prompted the regional authorities to develop information technology. As a region of a Nordic welfare state, North Karelia aimed to take care of its inhabitants, to include them all in societal development. There were a number of ongoing development projects focusing specifically on citizenship and the information society in the area. Many of the development projects were partially funded by European Union structural funds for peripheral areas. One of the projects, North Karelia Towards Information Society (NOKIS), participated from 1997 to 2001 in a network of twenty-two European regional projects of local information society development. It was based in the North Karelian regional council and it developed the regional information strategy. It also encouraged people and organizations to build up development projects from the local citizens' starting point, as an alternative to development done in technical terms. Local funding agents, such as those which distributed EU funds locally, used the regional local agency strategy as one of their points of reference when they made decisions. Women's information technology projects which aimed to empower women's agency in a technologically mediated society, to be discussed here, fitted well into the local strategy.

Together with Sari Tuuva and Johanna Uotinen, I followed several projects related to information society development from 1997 to 2000. We interviewed women participants, teachers and developers of the women's information technology project, attended their meetings and collected documents and media clips related to it. I use this process as a starting point for studying the construction of local agency by relating it to the texts of information society strategies on national and regional levels, as well as to the organizational practices of the women's groups. I study authorities' definitions of the information society (in the strategy texts), as well as the organizational relations in and participants' interpretations of the project. I also map the social and textual practices, including the social orders embedded within them (Smith 1990; also Fairclough 1995), that intertwine with the gendered situated local agency (Haraway 1991) in the North Karelian province in a process that the local actors call information society development. The starting point of the analysis is in the definitions of the North Karelian inhabitants, developers and women involved in the development projects.

Information technology in women's groups

The first women's group in my case study met together in the neighbourhood centre in a suburb of the regional capital, Joensuu (Vehviläinen 2001a). The women had various backgrounds, from busy working women to students and mothers with part-time work, and they all lived in the suburb. During the summer they made labels for canning bottles; in December they produced Christmas cards. They learnt to do letters and files using the computer, and they used the Internet to search for information on, for example, childcare and cooking, as well as their work activities.

Sari (not her real name), aged 36 and the mother of three young children, was an active member of the group. She tells about her participation in an interview:

> I work (sell cosmetics, study for a new occupation) while the children sleep ... We have a computer at home, bought by my husband, for example for games for the girls. I did not, however, at first see how I could use it myself, but it was first in the group that I learnt to see my needs ... I was asked: what are you doing there? Have you used computers? I had not been using the computer, and then other women also felt that they could join us although they had no experience ... Each woman was able to say what she wanted to be done. The group first met every fortnight ... In the group we speak about other things, too.

This women's group was part of women's own grassroots activity and it approached information technology from the starting point of their everyday lives, work and leisure. The group as such was important. It allowed women to change views of their neighbourhood and it allowed them to approach information technology from their own starting points. The neighbourhood women's group, although temporal and gathered for a specific purpose of developing information technology, intertwined with women's agency. On this basis, women themselves were active and made information technology another layer of their agency, in terms of their own lives and situations. The group gained media attention and provided a model for further development and the involvement of North Karelian women.

After some negotiations with the NOKIS project, a large institute for vocational education took up the idea and ran a partially EU-funded project of women's information technology groups in 1998 and 1999. Six groups of ten to twelve women gathered together with female teachers, in rural areas as well as in the regional capital, over a period of a year and a half. Furthermore, the women participants were invited to the institute for workshop days that were taught by male teachers. The project had a full-time manager who supervised the teaching and development, a project group for the discussion between the teachers and the manager, and a steering group. The project manager mediated between the groups. As a result, the women participants took standard exams in basic user skills in computing. Women's grass-root activity in developing information technology of their own turned into information technology education.

Developers in dialogue with women participants

There are various people on the scene when we are talking about concrete projects. All of them act from the starting point of their situated and historically shaped life settings. Their agency is constructed within the various orders to be discussed here, but they also make varying interpretations and definitions of their own in the context of their lives. They all deserve space in my narrative. There is, however, one particularly central figure in the process, Tuula Ikonen. She first initiated the women's information technology groups and later pushed for a project on women's information technology groups. She, too, acted in the project steering group and participated in one of the groups as a student. She is one of the few women in North Karelia who has contributed to long-term development in women's politics. She invited me to a dialogue during the planning phase of the educational project, and here I continue the dialogue further.

Tuula Ikonen (2001) has a political background from both the municipal and national levels; she has served, for example, as a member of parliament and has been a member of the Centre Party, which has brought the perspective of farmers and rural areas more generally to Finnish politics. During the period in which she was in parliament she promoted gender equality issues. During the project, she was no longer involved in active politics but worked to promote social and health services in a project funded by the European Union. She faced all male information technology groups in her neighbourhood centre, and she reflected on the situation in terms of women's politics. She wanted to make space for women's practices of information technology, and she called for the establishing of a local women's information technology group. Later she wanted to make the successful women's group a model for further development:

> There is a danger that women are displaced in information society development. They participate less than men in the construction of computer programmes etc. During the 1990s the field has become even more male dominated ... I made a first draft to continue the project for the neighbourhood women to get funding for their education and for inviting others.
>
> (Ikonen 2001: 151–2)

As a former member of parliament, she knew the people to network with to forward the idea for a bigger project:

> Easy networking between groups is one of the strengths in North Karelia: the regional government, the regional council, a development funding agent, educational organizations, associations, the third sector, etc. Also projects: women's projects and others have networked with each other.
>
> (Ikonen 2001: 152)

She invited me to articulate women's information technology politics and I helped her to write some drafts of the project plan.

Within North Karelian actor networks, the project was then allocated to a big educational institute. Tuula Ikonen attended meetings in the planning phase and in the steering group and explained the importance of women's own information technology that is based on their everyday lives and discussed in women's groups. She attended one of the women's small groups all through the process. At the end, after a large regional information society seminar, she comments:

> There were amazingly many participants in the Information network seminar, and more than three hundred followed it through the Internet. I was surprised that the majority of the participants were women: entrepreneurs, women from working life, from rural areas and from cities.
>
> (Ikonen 2001: 154)

Although the project process went rather differently from her planning – i.e. women's everyday life as a starting point for women's own active agency was rather undermined in the second stage of the project – she continued to work in the project. The particularities of the development projects seemed to be of secondary importance for her. They needed to be taken seriously, but still it was perhaps most important to keep the development projects going. The projects give space for people's active agency and, amazingly, at the end there was a large group of active women. They were not the women who participated in the particular project but they were North Karelian women, and that was a positive development for the province.

Women who took part in the second phase of the project were not aware of the ideas about the everyday starting points and the early planning of Tuula Ikonen. They were offered free information technology training in women's groups near their homes: "From the very start the title indicated that it was meant for rural women, and of course this was the first indication that nothing could suit me better. It was saying that it was my turn now" (Tuuva and Uotinen 1999: 209).

If they managed to follow the teaching schedule, they were usually content to study with other women with a woman teacher. They passed their exams and many of them bought computers or new programs. The project even helped in installing them. They intertwined information technology with their everyday lives as well, but there is a difference compared to the first group. In the second groups, the technology was studied in terms of standard educational packages and it was not discussed in terms of the situated being of women. There is more to be said, but let us first turn to the textualities of the information society which define the forms of rights and citizenship, in order to have more tools for the discussion.

Textual orders: information strategy texts shape agency

The development of the Finnish information society has been co-ordinated through a number of strategy documents; for example, ministries have made strategies for their areas of authority. Among the strategies, the one entitled *Quality of Life, Knowledge and Competitiveness* (1998) aims to be a national, common-to-all strategy for Finnish information society. It defines the space of agency in Finnish society, and one of its aims is equality: "to provide equal opportunities for the acquisition and management of information and for the development of knowledge" and "to strengthen democracy and opportunities for social influence" (p. 10). The equal opportunities are provided through making services and basic skills available to everybody and by providing easy-to-use products. After this is done, "each individual, business and organization has responsibility for developing their know-how and availing themselves of the opportunities offered" (p. 11). The strategy defines Finland as a forerunner in information society development in rather technical terms:

> In terms of international comparison, Finland is at the forefront of information society development. It is estimated that two out of three Finns use information technology in their line of work, one in two has a mobile phone [this was in 1997, by 2002 the numbers were much higher], and one in three has already used services available on the Internet ... The information society can be called a network of interaction between individuals and information network.
>
> (p. 8)

People are assumed to be ready for such opportunities and new skills

> as citizens, consumers and employees. The skills are gained in everyday interaction and in work tasks, as well as in training offered in different sectors. The decisive factor is the individual's own initiative. Different communities must support their members in seizing the opportunities available and developing their competences.
>
> (p. 11)

Furthermore, there are groups of people that need to be taken care of: "those who have not had instructions in the new skills during their education or in their line of work" (p. 18). On the other hand, "young people's growing knowledge and skills stand in good stead in the development of such services" (p. 20).

The Finnish information society is built by individuals who make initiatives of their own regardless of their place in societal relations. Some groups are mentioned – the elderly and the displaced – in terms of basic skills, and those groups need to be taken care of. Finnish society still defines itself as a welfare state, which looks after those who cannot take the initiative themselves. Yet the active agents who build and shape the society are individuals. They are citizens in terms of liberal theory, citizens who have universal rights to equal access and who are able to make decisions of their own apart from their daily practices and social relations of gender and class that organize their lives. The only societal difference arising is the difference in basic skills of technology and information use, in the use of the opportunities provided by the development of the information society. This is a gendered difference, but this is not reflected in the strategy.

When societal orders like gender are ignored, this also means that people's ability to take advantage of the opportunities meant for everyone is left unexamined. People live situated lives, and equal opportunity is not enough for most of them (e.g. Henwood *et al.* 2000) to use the opportunity in their own ways. The strategy does not elaborate on the problematic of practices and societal relations of citizenship, but rather co-ordinates the development of the Finnish information society and, for example, funding, in terms of services and skills related to information and communication technologies. Those services and skills are something to be proud of, even by international comparison; they are to be arranged for everyone, and thus for women, if they wish to study them in groups. Women's information technology groups should be funded if they aim to acquire basic skills in ICTs, but the starting points of women's groups are not included in considerations. The situated active agency of women that acknowledges societal differences gets no support from the strategy.

In contrast to the national strategy, the North Karelian information society strategy for the years 1999–2006, *By Joint Work Party to the Information Society* (1999), emphasizes the citizens' perspective:

> So, the information society is coming – but in whose terms? When people talk about the development, they are often enthusiastic over technology, but, still, we are talking about a society for human beings, which is meeting new challenges and a new kind of action environment caused by technical development … In the citizens' perspective the people themselves affect what the information society is like in which they will act and how. Then technology will serve as helper and is shaped by its users. The information society is not only an isolated sector just for professionals but a natural and useful part of people's everyday life, leisure and work … The information society is an entity of interactive communities built by North Karelians, in which information technology is exploited for one's own needs. Not using the information technology must not lead to displacement, but basic services and rights have to be guaranteed for everyone.
>
> (pp. 1, 5)

> In order to avoid the, from our perspective, harmful development coming from the outside, we ourselves have to launch a process, with which we have an influence on what kind of a information society we want. It is not determined by technology and technologists, but a user-oriented information society rising from the objectives of the inhabitants of the area … North Karelia has all the qualities to act as an experimental field for an information society in which technology has been developed and experimented in real using situations for the needs of its users. In a user-oriented information society it is not enough that the latest information technology is physically present. Mere Internet, efficient mobile phone and communication networks, modern micro computers and information technological know-how will not yet create a user-oriented information society. User-orientation is born then when this technology is being exploited and applied to the individual purposes of the users, the North Karelians, from their own interests. Technology and equipment will be then easy to use and applications worked out with people. Information technology has to respond to the needs of very

different citizen groups and it has to be available to all groups ... Instead of receiving technology passively they themselves have to be able to influence their own future information society.

(pp. 17–18)

The North Karelian strategy emphasizes the citizens' perspective and people's own active agency as developers and shapers of the information society. It is primarily the society, communities and people, as well as people's ability to act from their own situated starting points, that matter, not simply technology and the skills in using it, as understood in the technological determinism present in Finnish discourse generally. At least on some level, the strategy follows the social shaping of technology and actor network approaches (MacKenzie and Wajcman 1985; Grint and Woolgar 1997; Law and Hassard 1999). The then executive director of the North Karelia regional council, Tarja Cronberg, has previously worked as a researcher of technology and everyday life in Denmark and Sweden, and as an expert in the EU network on the social shaping of technology; she was elected in 2003 as an MP. The strategy has been developed within her office, and she has been able to supervise the articulation of the local and situated agency of people in strategy writing. At least at the level of principles, the North Karelian strategy supports situated citizenship and agency, unlike the national strategy.

The North Karelian media have given space for the development projects that follow the citizens' perspective approach. The ideas, maybe most clearly articulated by Tarja Cronberg (1999), are rather familiar in the region. There have been parallel projects on citizens' own agency going on in the region, for example, to revive the area of high unemployment. The information society strategy only shifts the focus to the information society. The first women's information technology group appeared as an ideal type of the North Karelian strategy and it was often presented in local newspapers and radio. The citizens' perspective was supported by the funding agencies, too; for example, the locally delivered EU structural funds for information society development are not distributed only along the lines of the national strategy but also in terms of the local emphasis on the citizens' perspective. It was relatively easy to find funding for the expansion of the women's information technology group project.

Although the large education institute started a women's information technology project by referring to the regional strategy formulations of agency in the information society, it later followed the example of the national strategy. The institute educated women through standard ICT training packages and the training was provided in a top-down manner. This method does not support people's own active agency and definitions of one's own as intended in the North Karelian strategy. Instead, it presents opportunities for acquiring basic skills as the national strategy suggests. Along the same lines, women studied to obtain the basic skills as individuals, faced the opportunity as individuals as written in the national strategy. Although the teaching was done for women's small groups, the group community was not used to develop the contents of studying: the group served only as a form of organizing students.

Women in the second groups faced technology as individuals, and as individuals they related it to their everyday lives. If the first women's group was an ideal type of the North Karelian regional information society strategy, the second groups were an ideal type of the national strategy (Vehviläinen 2002). Citizens were provided with an opportunity to gain information technology skills, and they indeed took the responsibility and used the opportunity.

The second groups illuminate the drawbacks in the understanding of liberal citizenship in the information society. There was one woman in particular – let us call her Irma – in one of the groups who did not know how to type. She was a very active woman and she took responsibility for many other things, but the lack of typing skills made it impossible for her to continue:

> I am not able to learn as fast as those people who have used computers for many years, and I am completely new. [It should be for beginners?] Yes, it was, but now we are starting the phase with those who have proceeded a bit more, and it is natural that they feel, if they need to pull somebody behind them, that they are not able to proceed fast enough, and I cannot go along with them. I must go slower or drop out.

And indeed she dropped out. The project meant for everyone, among women, displaced her. The understanding of equal opportunity in the national strategy does not work if it makes active people feel guilty for not being able to follow the teaching meant for everybody. Citizenship and active agency need to be reflected by taking into account societal differences, taking part in people's agency. It cannot be the responsibility of each individual to take up the universal opportunity meant for everybody, but the opportunity should be situated as well.

There exist simultaneously various strategy and other "governing" texts, and people, organizations and other actors use them for their purposes and in their situated contexts. They legitimate their particular activities and, for example, apply and get funding for their activities by referring to these texts. They read them and organize their activities accordingly. In North Karelia, both national and regional strategies live in parallel; the same actors in a project can refer to both, situation by situation. Thus alternative strategy texts do matter. They cannot change the whole system nor all people's practices, but they do provide space for alternative activity for people.

Social, material orders: institutional practices

Texts, strategies and technical artefacts matter, but they matter only when people interpret and intertwine them with their local and situated practices, which are organized through social and material orders. In the planning phase it is relatively easy to refer (at least partially) to a new and alternative text. In actual work, organizational and institutional practices, with their histories and conventions, comprise an important layer in local agency. They can both support and restrict one's own active agency.

The first women's information technology group gathered in a neighbour-hood centre, which was supported by a (partially EU-funded) welfare network of residents and experts: social workers, city planners, youth workers and others. The network discussed co-operatively the well-being of the neighbourhood and, for example, took care of the centre and arranged equipment for the inhabitants. The social workers supported the residents by discussing, encouraging and giving advice for their self-help and practical arrangements, but they never arranged things *for* people. They used the services of the welfare state to facilitate people's grass-root activity, the local agency (Kinnunen *et al.* 1998). They were partially financed by the European Union but they were staff members of a Nordic welfare state. The first women's information technology group was initiated by an active woman resident. It gathered women and developed women's own goals based on their starting points. The welfare network did not start or run it, but it provided the basis from which the group could start up and get going.

The second-stage women's information technology groups were arranged by a large educational institute. The North Karelia Towards Information Society (NOKIS) project negotiated with the institute and an EU structural funding agent on a project to develop educational practices in the citizens' information society, along the lines of the North Karelian information strategy. To some extent it also helped in project planning, but soon the educational institute took the responsibility and designed and ran the project.

The new project turned into a rather conventional Weberian hierarchical organization. Women's groups were taught by women teachers, but the teaching consisted of standard basic skills in information technology and women students took standard exams for "ICT passports". The women's information technology project was only one project among numerous others for the large institute, and although there was an interest in developing new educational practices, the institute could not make major efforts to develop its organiza-tional practices in order to support the new goals. The hierarchical relations typical of large public organizations intertwined with the women's information technology project and shaped the group process towards educational practices which provided people with basic skills, along the lines of the national informa-tion strategy. These organizational practices did not support women's situated learning and agency.

In some cases the group started to act as a group and women started to discuss and help each other. In those groups, information technology was also discussed in the context of women's lives. It was still taught as packages by the teacher, but women started to evaluate it by commenting on it to each other. This was not systematically supported, but some teachers knew how to leave room for women's own reflections. Even within hierarchical practices there was some space for situated definitions and the agency of women. They started to build situated agency. The practices between the teacher and the small group comprise another important layer in local agency, and are a layer where the nexus of gender, expertise and information technology can be challenged at the level of active subjectivity.

Organizational practices are difficult to change but yet they can be reflected on and developed. There are numerous examples of critical and feminist pedagogy (e.g. Weiler 1988), which give space and support for students and student groups to develop expertise of their own, and these examples can be used in large educational institutes, too, for small group processes. The welfare network that supported the first women's information technology group had made the effort to reflect on its practices, and it managed to change them, to have dialogue among many instead of using conventional hierarchy. The organizational and institutional practices are part of local agency and the reading of societal texts in actual practices, but they are not deterministic forces. They can be challenged and developed. If not reflected on, they (too) easily restrict rather than support situated local agency.

Gender seems to be self-evidently present in a women's information technology project. When men are the norm in the expertise of information technology, it is women who represent the visible gender. On the one hand, this happened in both groups; on the other hand, the understanding of the gendering process was very different in the two cases.

In the case of the first group, the women together made an alternative women's information technology group. They studied and developed information technology from the starting point of their lives as women who inhabit the particular region of North Karelia. They developed both practices and contents of information technology in a manner that made space for gendered being and acting, for their gendered local agency. While men's groups examined programs and networks as such, as if technology were shaped distinct from the rest of the social world, the women made labels for jam jars and other things related to the particular time and space of their lives. In this way, they truly challenged the realm of the ungrounded technology and its nexus to male expertise and made their own mark there.

Women in the second stage of the project continued to study in a small women's group with a woman teacher. Now, however, they studied the basic skills of information technology as given to everybody involved with the educational institute. The women's groups did not challenge the content of information technology from a gender perspective. Yet the women's group itself turned out to be good practice for women's learning. In all cases the women participants expressed the fact that they liked to study in a women's group. They felt that they were able to ask questions better than in mixed groups. A women's study group is a forum for some empowerment and redefinition of the relations of gender and expertise in information technology. In women's groups women can study on their own level, while in mixed groups they (may) feel stupid, which certainly restricts their learning.

Sometimes women's groups and their teachers gathered together at the institute for workshop days. Then they were taught by male teachers. This way the educational institute signalled that there is something in information technology education that cannot be taught by women, that there had to be male teachers for some areas of expertise. It seemed to state that the expertise of information technology is highly gendered: there are areas where women teachers can teach, but there are also other areas where women teachers cannot be used. There are some things beyond the reach of women, where only men can be experts. This split reflects closely the Western stereotypical understandings of gender and technology relations; it is passed unnoticed and is seen as self-evident (one teacher takes this up in an interview). It strengthened rather than challenged the gender relations in this case, and in this way a women's information technology group project engaged in unfeminist politics.

There has been little discussion about the construction of gender equality in Finland. Instead, there is a strong rhetoric that considers equality between women and men as an already existing fact – although the segregation in the labour market and in domestic work is exceptionally deep and a wage gap between women and men exists (Rantalaiho and Heiskanen 1997). Similarly, there has been little discussion about the gendered (or, in fact, any socially constructed) bias in information technology and the fact that there are very few women's activities or politics within technology, although gendered orders shape people's access to technologies (e.g. Compaine 2001). The women's information groups discussed here did not have Finnish examples to refer to, and they needed to struggle their way forward in a rather undefined terrain with relatively undeveloped (compared, for example, to the UK or even other Nordic countries) concepts and terms. They managed to produce a contradiction in terms.

The welfare state and organizational practices intertwined with gender relations serve here as examples of the social and material orders that interact with local agency in a technically mediated society. There are other relations or orders as well that are all present in local agency: class relations; age; place, together with the labour market related to the particular area and distances; relations between transnational, national and regional agencies having a concrete impact, for example, in funding processes; technology push; community-based relations; and societal gendered relations of income and the division of labour, in both public and domestic spheres. It is in the concrete practices that these relations take place.

Finally …

In this chapter I have discussed the local agency of people in a technologically mediated society within the textual and social orders of the society, as well as in people's situated subjectivity. People and organizations act within textual orders in their specific social situations and also use them for their particular purposes. The developers, as Tuula Ikonen in the project presented here, may consider the development as such important for active agency (for the development of the region or other large societal programmes) and look for suitable concrete projects and references within the national and regional textualities and alliances among the institutions and other actors. The local inhabitants intertwine technologies with their everyday lives. In local communities or small groups they are able to reflect on this process and can approach technology from the starting point of their local situations. Institutions and organizations contribute layers of their practices to people's agency. In a hierarchical organization the actors are treated from above, by command, and they are made to learn particular things within a particular timetable. They are supported in an agency that adapts itself to the technology push of the surrounding society, instead of helping them to evaluate it. Institutions and organizations should also reflect on their own practices in order to participate in the citizens' information society or in the processes in which actors shape the technologically mediated society.

Local agency in technically mediated society is more than "equal access" to individual (lonely hero) citizens. It is shaped in a broader societal and institutional context. One's own articulations, experiences, situations and definitions have space only if institutional structures and practices are built to support them. Institutional support is important for all, but it is especially important to those who depart from the (liberal) ideal of a Western information society: a middle-class, well-educated man. Different and even contradictory figures, like a woman who cannot type, should have room in the courses developed in terms of "equal access". The welfare network provides an alternative approach. The institutional actors can reflect on their own agency and can develop it to give support and space to people's agency. This is a far from self-evident process. Even grass-root activity easily turns into the forms and practices of hierarchical institutions, and rural women get trained towards the ideal of a universal cyber-cowboy whose only meaningful difference is in technical skills.

The first women's group respected women's own situations. They did not discuss the male gender, technology and expertise nexus, but they developed ICT of their own. The women's groups in the institutional project, in contrast, gathered in the spirit of a universal, equal and liberal citizenship. Gender was not seen to have any effect on the contents of ICT. ICT was understood as neutral and similar for everybody. Women's gender was defined only in terms of social interaction: it was nice and safe for women to study ICT in women's groups. This understanding of gender implied a result which was opposite to the original goals. In some areas the women teachers were not dealt with as competent experts but were replaced by men. The project strengthened the mainstream gender division between men as experts and women as users.

Gender and expertise relations are defined in everyday practices, and these provide possibilities for change. The project experience suggests that the relationship of gender and technology needs to be articulated openly. If unarticulated, the deep bond between male expertise and technology may appear as self-evident, and such projects can, paradoxically, turn against their own goals, limiting women's agency rather than encouraging it.

Women's ICT groups as such seem to be a good forum for women to build the agency of their own within the tricky gender relations of the technically mediated society. In order to create voices of their own, women need to connect male technology to their own starting points, and that means struggle and work. Their own starting points need to be defined and articulated, and a women's group can provide good support for this. People's own starting points are constructed in complex societal relations, for example around career prospects or available money, that often intertwine with gender. The groups that start from women's situations and everyday starting points can together make situated technology and have room for societal differences and ICT developed from the starting point of the differences. The groups have room for differences between women and in women, but they also make space for a communal agency that shapes structural and cultural relations through and within individual choices.

4 Empty spaces without knowledge and management

Riitta Kuusinen

In the information society information is gathered and modified in and through numerous different projects, both within and moreover between work organizations. Various Intranets and the Internet provide novel possibilities for project-like information processing. Boundaries between individual work communities are becoming blurred. More or less permanent groups and new ways of working, that exploit new technology, are being established for information processing.

Changes in working life, or in modes of action in shared information processing, are not easily made: problems around this occur. The attempt to control the social phenomenon of information processing has resulted in new kinds of challenges, particularly for knowledge management in work organizations. Knowledge, however, is an undefined subject to control; for a start, the social phenomena related to it are not visible to the naked eye. Suitable approaches, concepts and theoretical models for these aspects of working life are being increasingly explored in a range of disciplines.

This chapter draws on an action research project examining modes of web-assisted information processing. In its experimental phase (1996–7), the project, in which I have been the key researcher, involved the creation of a web forum for which, according to the aims of the funder (The European Social Fund in Finland: ESF–FIN), about seventy ongoing projects (all funded by ESF) in different parts of Finland carrying out research on changes in working life were to produce interactive knowledge.

Policy-makers and developers of education and working life were thus to receive for their use up-to-date "anticipatory information" about current changes in working life practices. The main forum of the overall "anticipatory information" project was a public database that could be used on the Internet. Its purpose was to become a shared public forum with the aim of conveying the latest information and knowledge for policy-makers and others that might benefit from knowledge regarding future needs in working life. This web forum was named ENTTU – an abbreviation of the Finnish *ennakoiva tutkimus*, in English "anticipatory research".

The web forum, however, remained an empty space despite the numerous different attempts to support and encourage its use. In my capacity as the manager of the web forum, I asked for support in the form of substance-based knowledge management among projects. This was not provided, nor was there any willingness to provide it in traditional public administration (Ministry of Labour, Ministry of Education and Ministry of Social Affairs and Health/ESF–FIN).

After the experimental phase finished, I analysed the factors that regulate shared information/knowledge processing on the basis of the data and a range of theoretical contributions. In 2001 I published my doctoral dissertation on the subject, which presents my interpretation of the theoretical foundations for web-assisted co-operation in information processing (Kuusinen 2001).

In this chapter I re-examine the question of the possibilities for shared knowledge in relation to this research project. This has led to the development of a theoretical framework for information/knowledge processing, as seen mainly from the knowledge management perspective and especially from the viewpoint of human knowledge and the organization of shared knowledge processing. These issues are extremely important for the everyday practices in working life within the developing information society.

Interpretations of knowledge

The interdependency of knowledge, the human being, the community and the organization has been difficult to define; scientific consensus on this has not been reached. The human being is recognized as central in knowledge-intensive work but there are many, often contradictory, conceptions of knowledge and the impact of the environment upon knowledge.

The Finnish researcher Ilkka Tuomi (1999) has recently provided an extensive review of the literature on knowledge management and categorized the publications according to their emphasis. In those emphasizing the human being as a resource (e.g. Sveiby and Risling 1987; Stewart 1997; Brooking 1999), Tuomi has observed a recurring notion that human capital and structural capital are two different types of knowledge resources. He illustrates this by an example: human competences "walk out the door every night", whereas structural capital "stays in the company". Tuomi himself is of the opinion that individual competences exist only in relation to organizational systems of activity. Thus human capital does not "walk out the door"; instead, people go home and their competences remain within the organized system of activity. This activity, however, is interrupted when people go home: motives that relate to organizational activity become latent. Tuomi thus sees individual abilities at the workplace as closely connected with the system of organization.

Carl Bereiter (2000) also considers conceptualizations of knowledge that are limited to it being seen as the possession of individuals to be too narrow. He cautions against treating human beings as if they were simply knowledge containers. He speaks about the mind-as-container metaphor and reminds us that the mind is only metaphorically a container. The metaphor has to be approached with caution because it perpetuates an inaccurate conceptualization of knowledge; in other words, knowledge is easily understood as mental objects which can be transferred into the human being's knowledge container through education. Correspondingly, learning is seen as consisting of taking objects in from the outside. This is contradictory to the widely accepted conception that human beings construct their knowledge themselves and that education acknowledges ideas such as higher-order skills, teaching for understanding, authentic problem-solving and lifelong learning. According to Bereiter, there are too few unambiguous words to meet the needs of the information society. Terms like "learning", "information" and "knowledge" remain vague. In addition, various professional groups use their own concepts. Nowadays there is a need for increased inter-group co-operation, however, and that is when the meanings of concepts become problematic.

Bereiter argues that knowledge has a dual character. On the one hand, it is something real in itself: for example, the knowledge contained in books, the state of knowledge in a field, the advance of knowledge, the sharing and dissemination of knowledge. On the other hand, knowledge can be considered as a form of content in individual minds, like acquiring knowledge, or having much or little knowledge of the subject. He has found that when people are writing articles for prestigious journals they avoid talking about knowledge as having a life of its own and stick to the things-in-the-mind conception of knowledge, believing that to be the more easily defensible one.

Bereiter considers it dangerous not to care about the human elements of knowledge, especially when the point of deliberations is to reform human institutions such as industry and schools. He presents two approaches to resolving the dual character of knowledge. One is that knowledge is constituted in communities of practice and embodied in the tools of such practice. In the alternative approach, knowledge can be seen as conceptual artefacts, which are immaterial and serve purposes such as explaining and predicting.

Ikujiro Nonaka and Hirotaka Takeuchi (1995) have discussed how individual knowledge can be brought into a company in order to increase its competitiveness. They lean strongly on Polanyi's (1966) concept of tacit knowledge. According to Nonaka and Takeuchi, the success factor of Japanese companies has been the appreciation of the tacit knowledge that their employees have gained during their work experience and that includes subjective insights, intuitions and hunches, as well as ideals and emotions. Employees' reserve of tacit knowledge is the main resource in the creation of organizational knowledge. By that Nonaka and Takeuchi mean the capability of a company as a whole to create new knowledge, to disseminate it throughout the organization, and to embody it in products, services and systems. Knowledge creation requires frequent intensive and laborious interaction among the members of the organization. Through discussions, tacit knowledge can be modified into explicit knowledge and further into new products and practices of the organization. Explicit knowledge is "codified" knowledge, referring to knowledge that is transmittable in formal, systematic language.

Nonaka and Takeuchi emphasize the human elements in organizational knowledge creation: an organization cannot create knowledge without individuals. The organization's task is to support creative individuals or to provide contexts in which to create knowledge. They talk about a process that "organizationally" amplifies the knowledge created by individuals and crystallizes it as a part of the knowledge network of the organization. Their dynamic model of knowledge creation is based on the assumption that human knowledge is created and expanded through social interaction between tacit knowledge and explicit knowledge. They call such interaction "knowledge conversion" and emphasize that it is a social process between individuals and not confined within an individual. Nonaka and Takeuchi present four modes of knowledge conversion that are created when tacit and explicit knowledge interact with each other: socialization, externalization, combination and internalization. Their perspective thus emphasizes increasing organizational knowledge through individuals and their actions.

Jean Lave and Etienne Wenger (1991) in turn look at how individual knowledge accumulates with the help of a community. They have brought to the fore the significance of community in individual learning by using the concept of legitimate peripheral participation. By this they mean that learning is itself an evolving form of membership. Participation in social practice suggests a very explicit focus on the person, but as a person-in-the-world, as a member of a socio-cultural community. This focus promotes a view of knowing as activity by specific people in specific circumstances. From this perspective, learning involves the whole person; it implies a relation to social communities, it implies becoming a full participant, a member, a certain kind of person, including the construction of identities of personhood.

Lave and Wenger (1991) criticize earlier theories that, for example, address its socio-cultural character by considering only its immediate context. They argue that in the social world it is important to consider how shared cultural systems of meaning and political-economic structuring are interrelated and how they help to co-constitute learning in communities of practice. In the concept of legitimate peripheral participation Lave and Wenger see two important elements: first, the development of knowledgeably skilled identities in practice; and second, the reproduction and transformation of communities of practice. Thus, in their opinion, the concept of legitimate peripheral participation can act as a conceptual bridge, referring to the common processes inherent in the production of the changing of people and communities of practice. Lave and Wenger define a community of practice as a set of relations among persons, activity and the world. It is an intrinsic condition for the existence of knowledge – it provides the interpretive support necessary for making sense of its traditions and past development. The social structure of the practice defines possibilities for learning.

No shared activity without organization

In this overview so far I have presented some of the more well-known examples of the current scientific discussion in which notions of knowledge, individual and community converge. The sharing of organization and the organizing of shared activity are often ignored, even though they are essential in knowledge management. Shared information/knowledge processing through a new medium is shared activity and it too requires organizing; indeed, this feature became evident in the experimental phase of my project.

In an already existing organization it seems almost self-evident that the activity of the organization is maintained by the people who work there. They form a collective with smaller groups to attend to various tasks. Employees both get more knowledge and learn new skills that they can hardly use outside the organization. Organizational activity is not endangered even if a few people leave as newcomers then learn to master their tasks.

What if the entire community left the premises of the organization and a new, but existing, community from an entirely different field replaced it? The newcomers would find information lying on the shelves and stored in the computers, even though much or most of the information would be of little or no significance to the new community. They would probably empty these stocks of knowledge and start their own familiar routine, producing related documentable information for the same shelves and computers. Each of them would definitely take up their own former role in the shared "game" in these new premises and the community would soon be in full swing. In this example the community would be able to bring the learnt way of organizing their shared activity into the new premises as well; individuals would start acting by using the knowledge and playing the roles they had earlier learnt within and as a community.

Now, let us imagine a situation in which people who do not know each other beforehand enter an empty working space of a work community that has left the organization. A body that is significant to them has invited them to be there. Each of them actually works for another organization and participates in the activity maintained by its own work community. The ones who arrive would perhaps try to get acquainted by taking turns in telling the others where they come from and what they do. The sender of the invitation suggests that they use the working space for shared activity. Nevertheless, nobody feels the need to participate in the shared activity nor knows why they should try to develop such an activity. It can be assumed that after a few confusing moments they would become frustrated and probably leave to go to their home organization to continue working there. The available working space would remain empty. This imagined situation leads us to the question of what it is that would be needed to make a functioning community out of the group of people who had entered the scene.

The above example in this simplified form corresponds to the situation which emerged in the experimental phase of the overall project. The "working space" provided for shared activity of the "anticipatory information" project was indeed virtual, a shared web forum, which remained as an empty space without the input of knowledge that the participating projects were asked to produce interactively. This request was made possible by the funder, the European Social Fund in Finland (ESF–FIN), being the same for projects implemented in the different organizations.

It may be useful at this point to describe in a little more detail how the experimental activity turned out in light of my own experiences as a researcher, along with a questionnaire survey of the project members.

Results of the experimental activity

Project funders generally leave the individual projects to implement their research plans without trying to create interaction among them. The ESF funders, however, had vague expectations of creating a national anticipatory system, and a project was funded to co-ordinate it. Therefore, the representatives of funded individual projects started to receive e-mail messages in which the co-ordinator briefed them on various things. The recipient list on the screen consisted of strange abbreviations or strange names which enabled an individual project researcher to deduce that he or she belonged to some kind of group whose members shared the same source of funding.

The co-ordinator also arranged two-day seminars a couple of times a year. In them, project representatives presented their work to other participants. At the end of a seminar held on a ship, the chairperson announced that in order to increase interaction among projects, a new project had been established. It would maintain the web pages to be shared by all the projects. At that time, even home pages were rare and e-mail was used by few people. Therefore, mentioning such a project did not raise any concrete images in the minds of the project researchers, who then went back to their desks in different parts of Finland. Obviously the researchers had not been able to do their regular work during the seminar and therefore their projects had not gone forward; but usually the funder did not allow the research timetable to be too open. Nevertheless, the best policy was to take on the invitation when it was sent by the funding body.

The representatives of these anticipatory projects got used to including the reading of the co-ordinator's e-mail messages in their ESF-funded project work, although these did not actually further the goals set for the project. A few months after the seminar on the ship, a new project researcher sent a group e-mail message telling the others the address of a www database meant for the networking of the anticipatory projects, and advised everyone to use it. The projects then had a web forum in which to exchange information and ideas.

Along with the web forum, individual representatives of the anticipation projects were given new responsibilities. They had to write for the web forum and send information on their own project work: in other words, to present their own project. The strangest requirement was that their own incomplete thoughts would have to be written for others to see, which under the pressure of project work felt unreasonable. As a matter of fact, as the funding was awarded, the project researchers had committed themselves only to implementing their own research plans and producing their own final reports.

On the other hand, representatives of the anticipatory projects were interested in what kinds of subjects the others were thinking about – perhaps to get ideas for solving their own problems or even help by co-operating. Researchers, however, felt insecure and lonely in this community of people they did not know who were located in different parts of Finland. Their screens continually displayed messages concerning expectations and demands, but it was difficult to answer them just like that. The threshold of writing their own thoughts in public for everybody to see and to comment on seemed too high. Neither was their own work community of help in stepping over that threshold.

Central elements in shared knowledge management

The experimental phase of the overall project demonstrated that technology as such does not necessarily in practice produce shared knowledge processing. Technology can probably be exploited as an assistive means if a new kind of shared activity, knowledge-intensive collaboration or shared knowledge processing can be made to emerge in a group of people. I would like to stress that although I write at a general level, I am not writing about scientific truths but about my own theoretical perspective. In just one article it is not possible to present the relevant numerous experiences, justifications and theoretical interpretations that have led to the development of this theorizing. I hope that this theorizing can assist those readers who seek to understand the various similar phenomena of shared knowledge processing.

In the light of this experience, my own conception of knowledge is human-centred: only human beings can create and use conceptualized information. My conception of a human being is learning-centred: biological beings learn to become human beings in the social environment in which they grow. Learning how to become a human being includes both action and speech, and these basic means are stored in human knowledge processing throughout the lifespan. In addition, the capability to think enables human beings to be creative, to combine in new ways pieces of information absorbed from the outside and to apply them in problem situations.

Following this human-centred conception of knowledge, I seek to explain the phenomenon of shared knowledge processing in a human-centred manner. The basic elements of the phenomenon are as follows.

Organization represents a frame of action or the objectives, rules, financial arrangements, etc. that are created for a group of people. If an organization does not support people's activity in a desired way, it needs to look at what kind of codes of action and possible unconscious practices are prevalent in the organization.

Community is another basic element, which consists of several people. In a work organization, it is the community that represents the social environment which is part of the learning experience of its members. Action and use of speech provide individuals with common modes stored in the memories of individuals as shared knowledge.

In the situation of social knowledge management a group of people has been brought together on some basis to process information. It is the action strategy drawn up for a group of people, and in information processing particularly the *information processing strategy* that defines how the community processes information. In a traditional classroom, the strategy is individual-centred; to produce knowledge in co-operation, a group-centred strategy is needed. Thus the information processing strategy is crucial in creating new web-assisted information processing or when the methods of knowledge processing in a work organization require changes.

A group of people who share an activity also generally need a leader or leaders. Collective activity requires a *person(s) in charge*, who observes the activity from the viewpoint of the whole entity. In an equal and group-centred activity, the roles of the person(s) in charge are probably those of serving the community, creating possibilities to act, and co-ordinating the flow of knowledge. In such activity, the person in charge is subordinate to group decisions. The group needs to give the person in charge the duties and authority required by the task so that his/her actions would not be associated with a thirst for power. Because of his/her duties, the person in charge is in practice the leader that the group always needs in order to act efficiently.

One of the key tasks of knowledge management is to channel the individual knowledge construction accumulated within individuals into shared knowledge processing. In dealing with individual knowledge construction in a group, creative thinking can be realized by inspiring one another and by producing shared individual knowledge construction. Thus *group knowledge processing* constructs knowledge again by the same principles as for an individual in cognitive knowledge processing. Both select, interpret and modify available information. The group does this from the basis of individual knowledge construction brought in by individuals, estimating the significance of its parts from the viewpoint of the group objective.

In knowledge-intensive work, *individual knowledge construction* can thus be viewed as material to be dealt with. Various individuals construct their own knowledge construction even from the same sources – the same book or the same event, for example – selecting, interpreting and modifying it in their own individual way. A successful group discussion produces a knowledge entity that is more diverse than individual knowledge construction. At the same time, individuals learn shared knowledge, while also individually selecting, interpreting and modifying it in order to integrate it into their own internal stocks of knowledge, clothing it in their own conceptions, integrating it into their own understanding and re-evaluating their own activity according to it. Thus, even shared knowledge can hardly ever be exactly the same within various individuals, although substantive similarities can be detected in the fragments of their knowledge.

Individual knowledge construction

In seeking to theoretically understand the processing of knowledge in a group, my starting points have been individual knowledge and the central principles of its emergence. Figure 4.1 attempts to make it easier to understand this process of individual knowledge construction.

Individual knowledge construction is displayed here in the form of a triangle or pyramid, as human beings record their knowledge supported by a hierarchy of concepts. At the top lies abstract theoretical knowledge that is applicable to various practical situations. At the bottom of the triangle there is practice-related, experience-based knowledge. They are combined in a process of understanding resulting from individual thinking, in which individuals through their theory-in-individual-use look for explanations and solutions to the phenomena they have detected.

In cognitive information/knowledge processing human beings construct knowledge for themselves at all three levels, combining these closely together. Accumulations of knowledge represent learning. Learning that takes place in a conceptualized form I call *conceptual learning*, and knowledge accumulated by it *conceptual knowledge*. In general school education, teaching traditionally focuses on this area. Knowledge that is in the phase of being understood is still in search of its place in individual knowledge construction. It includes both conceptual and experience-based templates that are worked on when new stimuli appear. Learning-by-doing I call *action learning*, and the knowledge accumulated by it *action knowledge*. The social environment in which human beings operate defines action knowledge and is thus part of the formation of individual knowledge construction.

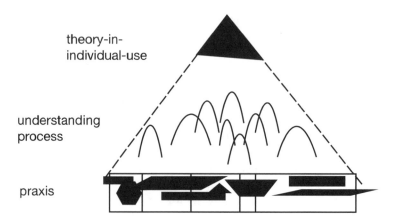

theory-in-
individual-use

understanding
process

praxis

Figure 4.1   Individual knowledge construction

Argyris and Schön (1975, 1978) have also raised the separateness of conceptualized knowledge and knowledge visible in action by using the concepts of espoused theories and theories-in-use. According to them, theories-in-use means action in practice, whereas espoused theories would seem to refer to the consciously perceived principles of one's own action models. In my usage, theory-in-individual-use rather refers to conceptual knowledge, and action knowledge refers to a habitual activity which is generally neither conceptualized nor entirely consciously perceived.

As early as the 1940s, Gilbert Ryle (1949/1963) brought to the fore two different fields of learning and knowledge: knowing how and knowing that. He justified this distinction by referring to the different ways they come to exist in the human mind. Human beings learn *how* by practice, schooled by criticism and example, but often quite unaided by any lessons in the theory. He describes that after initial learning, an activity becomes automatized so that it does not need to be thought about; it becomes the human being's "second nature" to do what is allowed and avoid what is forbidden. He emphasizes that these two different processes are not separate: efficient practice precedes theorizing of it.

A corresponding duality has been visible in recent psychological science as well, as in Lauren B. Resnick's (1996) work. She does not consider that apparently different schools, that she terms conceptual rationalism and situated cognition, contradict each other, but is of the opinion that they should be combined. Conceptual rationalism seeks biological foundations for specific concepts that are central and, perhaps, universal in human development. Situated cognition argues that knowledge is acquired in and attuned to specific social and historical situations, and that conceptual development can be understood only in terms of the situational contexts of action. According to Resnick, these schools share important epistemological assumptions and should be combined to provide a theory of cognitive development and functioning. She terms this kind of combined view situated rationalism.

Resnick (1996) argues that individuals develop their situated competences in each situation on the basis of their pre-existing structures, which have both biological and socio-cultural roots. Socialization in a culture begins at birth. In each situation, learning is the matter of beginning to act in the environment on the basis of particular affordances and possibilities of that environment. One's initial actions can be more or less successful, but they can also be partially successful, enough to keep one engaged but not enough to provide a ready-made set of actions. If they are partially successful, a process of attuning to affordances and possibilities in the environment takes place (Resnick 1996: 1, 43).

At the starting of a process of co-operation, group members are in a unique and temporary environment. They start to act on the basis of their individual understanding that includes their conceptualized knowledge, their way of understanding things and their modes of action they have learnt in corresponding situations. An important research subject is indeed how group members create and experience the affordances and possibilities of the environment, on the one hand, and attune to them, on the other, in a self-directing process of co-operation.

In dealing with the substance itself, group members are expected to present on the basis of their own understanding at least conceptualized information, and process it together in a group.

In the process of individual knowledge construction formulated above, we can thus find two different kinds of knowledge to which several researchers refer by using their own concepts: tacit/explicit knowledge (Polanyi 1966; Nonaka and Takeuchi 1995); procedural/declarative knowledge (Anderson 1996); knowledge constituted in communities of practice/as conceptual arte-facts (Tuomi 1999; Bereiter 2000; Lave and Wenger 1991); theories-in-use/espoused theories (Argyris and Schön 1975, 1978); knowing how/knowing that (Ryle 1949/1963); and situated cognition/conceptual ratio-nalism (Resnick 1996). In constructing my own theory-in-individual-use, I have combined these dualities according to principles of cognitive knowledge processing and named them as "action knowledge" and "conceptual knowl-edge", which individuals combine on the basis of their own thinking. Thus understanding process emphasizes the fundamental significance of one's own understanding, which is stressed by the constructivist conception of learning in the constructing of individual knowledge. Through one's own understanding, theory is connected with action, and based on the experiences gained in the course of action, a theory-in-individual-use is generalized for oneself, to be applied to corresponding situations.

Action learning in the core of new practices

From the standpoint of knowledge management, and especially within the information society, the particular principles of action learning should be known and consciously perceived. Action learning is apparently done mainly by the brain and the environment, unlike conceptual learning, which requires conscious memory work. Action learning is built on automatic action models, and these are triggered for use unconsciously, based on analogical, especially visual, information. This principle can easily be seen in learning physical performances, such as riding a bicycle, learning how to swim, and driving a car. These skills are first practised consciously, by using work memory in phases, and finally they are practised without consciously thinking about them, like self-evident modes of action, by letting substantive thinking draw on work memory.

Now let us consider the relation of this discussion to the experimental research on the "anticipatory information" project. First, it is important to consider the impact of the model that we are given in social information processing in school. Individuals in the classroom learn a self-centred model of information processing, in which other members of the community are rather points of comparison in considering their performances, and are thus also rivals. This model also includes the notions of authoritarian leadership and the associ-ated rights to assessment. If one's own understandings and the special features of one's own individual knowledge construction have not been appreciated in one's school education, one probably learns to keep quiet about them. One does not want to express one's own thoughts out loud, because only formalized knowledge is appreciated.

Team organizations in working life aim for group-centred knowledge processing that is based on the presentation of group members' own individual knowledge construction. However, thought models learned in school are later triggered in steering behaviour in working life. This is what happened in the experimental research project, according to my interpretation of the events. Despite their positive attitudes, members of the community did not dare to send to the public web forum texts that would have revealed their own incomplete understanding. The appreciation of formal knowledge, especially prevalent in the Finnish school system, is reflected in the following responses given by members of the community.

In the course of research I don't usually produce any formal interim reports.

I will send … material as soon as we finish the preliminary research report in the months ahead.

So it'll be enough to present the findings when they are in a presentable form.

In this phase, the findings of our project are still not in a presentable form. There is in my opinion nothing that would prevent the publishing of our report in ENTTU, when the time is right.

I'm going to send a summary of the final report.

From our viewpoint it is important to diffuse our research findings to a larger audience and ENTTU is probably a good forum for that.

It is very difficult to change automatic action models and the thought models crystallized from them. It requires the *unlearning* of an old action model and subjectively repeated actions which eventually produce a new action model. An automatic action model begins to break down only if it no longer leads to coping under somewhat changed conditions. At that point, one has to learn a new mode of action out of necessity. This kind of necessity for group-centred knowledge processing is created by means of knowledge processing strategy, organization and leadership. Through these means, individuals are "forced" to learn group-centred knowledge processing in a positive atmosphere. Furthermore, resistance to change is a natural principle in action learning. It is manifested in the persistent appearance of the former automatized action and thought models for a longer period of time, despite the fact that a new mode of action and thought is actively being learnt. In this phase, a supportive group and its awareness of shared learning are favourable to action learning.

According to my interpretations, in the research project (and especially its experimental phase), group-centred knowledge processing did not get support from either the organization or the leadership. The traditional, top-down mental background organization in government ministries did not steer the projects towards shared knowledge management. The ministries focused on project-specific supervision of financial matters. Their main aim was to have project-specific reports produced. However, the use of the web-assisted medium should already, in the planning phase of the action strategy, have been linked organizationally and socio-physically to the launching and maintaining of a real-time knowledge management system.

The co-ordinator's task was to lead the substantive processing, but this role of the leader by nature included mainly top-down briefing. There was no attempt to form a community or different kinds of inter-project groups. Representatives of the projects were thus not led towards interactive and group-centred knowledge management. In the experiment, individuals worked in their own home organizations and their experiences of the experimental community were only dispersed samples or excerpts on the screen, which did not provide enough support for the shared exchange of thoughts and ideas in public. From this, I concluded that in this kind of situation interactive knowledge processing also requires face-to-face meetings. Learning to process knowledge in a group-centred way is difficult even in a familiar work community; therefore, it might be too much to expect it to happen publicly without this.

Chris Argyris (1993) and his colleagues, in their extensive organizational development work, have also come up against the problematic of action learning. They have concluded that all organizations need two types of learning. Single-loop learning corrects errors by changing routine behaviour; it is thus incremental and adaptive. Double-loop learning corrects errors by examining the underlying values and policies of organization; as such, it requires leaders who continually model it and reward it. Argyris's research has involved an extremely large number of individuals in the organizations under study. He points out that while the researchers had expected variance in personality and style, almost all individuals seemed to have the same difficulties in dealing with double-loop learning problems. Therefore, the researchers arrived at the conclusion that there must be some causal factors common to these different human beings. They concluded that the common factor was the dominant or "master" programmes that individuals held on how to design their actions and implement their designs.

Argyris and his colleagues seem to have found the same phenomenon that I call the existence of automatic action models. Through this, people in the same organization develop a factor that regulates shared activity, in a way it is difficult to be conscious of and that is therefore also difficult to change. As an experienced researcher, Argyris describes well the problems involved in detecting this phenomenon: "It took us several years and several thousand cases to conclude that, for some reason, most of us have learned (early in life) a theory-in-use whose *effective* implementation requires actions that are *counterproductive* for double-loop learning" (Argyris 1993: 6–7, emphases in original).

Developing work organizations is very much about changing prevailing activities, which, at the individual level, requires action learning in particular. From the perspective of my own theorizing, when Argyris talks about the two types of learning, he actually means the different degrees of action learning. Single-loop learning involves the adaptation of an action model to conditions that have changed only a little. Double-loop learning corresponds to the learning of a new kind of action model. It also requires that individuals are conscious of the old and the new principles of action and that they deliberately learn the new, which in their opinion can also be promoted by education. Argyris neglects dealing with conceptual learning, which is realized according to the generally acknowledged constructivist learning principles and in the assimilation of conceptualized information.

Argyris also argues that successful action learning is linked with organizational arrangements. According to him, leadership education with a focus on theories-in-use and reasoning processes can be integrated with such disciplines as strategy, managerial economics, management accounting and management information systems, because these disciplines are themselves theories of action intended to help managers achieve specific goals. Argyris states that "If the intellectual structure, architecture, and reasoning processes required for leading-learning are the same as those embedded in the managerial disciplines, then integration will be much easier" (Argyris 1993: 7).

The experimental phase of the research did not provide the members of the community with enough organizational support for learning the new activity of co-operation in web-assisted knowledge production. As a researcher, I was thus not able to observe this kind of shared knowledge processing. Instead, the experiment led me to find explanations for problems in group processing of knowledge by drawing on other researchers' theoretical conceptualizations. From this research experience, I have worked on developing a behavioural theoretical foundation for future research.

Virtual co-operation organizations as a framework for shared knowledge processing

From the point of view of knowledge management, it is extremely important to gather research data on what kinds of practical measures should be taken to positively affect the success of knowledge processing within a group. My special interest now lies in web-assisted group processing of knowledge, because this has proved so difficult to realize in the experimental activity of my research project.

Some members of the community expressed their thoughts about the experienced difficulty in the following ways:

If I presented my texts in public, I'd like my subject matter to be convincing enough and that my text would be neat, appropriate, interesting, etc., and polishing that kind of text does take too much time.

It sometimes crossed my mind to ask about the subject ... I never sent my text, because I set my standards too high.

... that concerns my character [i.e. when I publish something in a large forum such as the World Wide Web, I want to be sure that what I present is "correct"].

When shared web-assisted knowledge management takes place within a work organization among employees who already know each other, it is possible that the necessary positive effects from the community have an impact even though people are not conscious of them or do not need to talk about them. A much more challenging situation arises when group members come from different organizations and do not know each other in advance. In this case, the elements of community that promote working have to be consciously created for the group, whose purpose is to jointly process individual knowledge construction.

In my understanding, no kind of process of co-operation is possible without organized shared activity. The web-assisted work of a group whose members are scattered in different places can be viewed as an activity taking place in a temporary virtual co-operation organization. Although the organization is virtual, people as actors in it are real. Therefore, knowledge management can be expected to take place in the virtual co-operation organization as well, on the condition that human cognitive knowledge processing is taken into account.

In organizing the activity of a group taking its first steps on the path of co-operation, I see the following two levels: background organization and group organization. The central regulating instrument in the background organization is the action strategy of the community, or what might be better called the "knowledge processing strategy". Social information processing needs such a strategy, although in traditional organizations such as schools people are neither conscious of nor talk about it. It has already self-evidently been internalized in the action models prevalent in the community. In these models, the increasing of group-centred knowledge processing should start from the renewing of the individual-centred knowledge processing strategies.

The body that draws up an information/knowledge processing strategy varies on the basis of the organizational structure behind each group and is usually linked with the responsibility for financial resources. Therefore, a teacher, for example, has no authority to change the central position of individual performance assessment in the school's knowledge management strategy. For a co-operating group acting as an intermediate between organizations, the knowledge processing strategy can appear as more or less clear expectations of the form of work performance and the time provided for it, or as expectations for reaching certain objectives. In any case, in order to process shared knowledge, a group-centred knowledge processing strategy is needed to create the concrete possibility of performing the required tasks. It is closely linked with the defining of the fixed basic objectives set for the group, on which the planning of the knowledge processing strategy often leans. For example, producing a certain kind of joint publication already as such defines the knowledge processing strategy required. This is again easily influenced by the individual information processing model assimilated at school, through which the easiest and most familiar solutions can be followed.

The group that is starting to co-operate is usually itself responsible for its own group organization; however, it has to make decisions within the framework created by its background organization. Shared activity requires organizing, which includes attending to various areas of responsibility. Although the group might seek to be self-directing, emphasizing that all its members are equal, it usually needs, as I see it, a person in charge to attend to practicalities. The traditional title of "group leader" easily gives a wrong impression of the tasks of the person in charge, central to which I see the servicing of the group in many different ways. This person in charge should apply a participatory leadership style and attend to the implementing of the opportunity for equal discussion. Group organization includes agreeing upon recording activities, and minutes of meeting decisions pull together group activity and are necessary in clearing the air in possible conflicts of interpretation.

The elements of community needed in information/knowledge processing can be produced through repeated meetings and by arranging face-to-face situations. The need for these was referred to by many community members in their responses:

> Somehow this thing will definitely start rolling, as soon as we with this questionnaire, for example, can get it started. As I said before, knowing people personally would help in exchanging opinions too.

> It [co-operation] does not really get air under its wings only through the media. Even in my own project I've experienced how significant personal contacts are. Talking about matters will take place in a totally new dimension when you're face-to-face with someone. In my opinion, in the networking of our projects and the launching of co-operation there is a threshold to step over. We should perhaps know each other better to be able to start exchanging ideas. It wouldn't probably be too much trouble; it would even sort of make your work easier. When you ponder things alone too much, your thoughts tend to start going around and around in your head.

> I particularly avoid circulating detailed information in public, especially because we all are so different.

> The purpose probably is, by writing, to discuss even shorter snatches of our thoughts, ideas, etc. for others to comment on. This kind of informal discussion through e-mail is probably a fairly new thing for many, especially because you discuss with strangers.

Organizationally, the knowledge process is shaped by periods of regular face-to-face meetings and virtual computer-to-computer meetings, during which knowledge processing can be implemented by means of information technology (IT). The web forum can only be an aid if this kind of process of knowledge processing can be created. The group will have to plan the uses of its web forum as a part of its overall plan for knowledge processing. The starting point is the knowledge processing strategy and the objectives to be implemented by group agreement.

During the face-to-face meeting, group discussion is the most important producer of community. In this experiment the aim was to construct interactive anticipatory knowledge: in other words, to combine the knowledge that individual researchers have by means of their self-steered interaction. This also required face-to-face situations. An opportunity to be together informally can give the members a chance to get to know each other better and lay the foundation for increased trust. In discussions about substance, the most important things are the appreciation and bringing to the fore of each member's individual knowledge construction. Shared ideas and constructing the shared knowledge require group discussion skills of everyone – skills that clearly have not usually been gained in education at school. In the spirit of mutual support and joint effort, the group can create a favourable atmosphere for compensating various missing skills. A good discussion can strongly influence the success of the knowledge management process. Discussion is also the medium through which symptoms of a latent dysfunction in the process are first revealed.

From the standpoint of cognitive information processing, the group is the responsible producer of shared individual knowledge construction. The group sets an aim for itself and, in order to implement it, selects, interprets and modifies meaningful knowledge. Its material consists of the knowledge that the group members present. Individuals need the community to define what piece of information in their stocks of knowledge could be meaningful to this particular group. At the same time, an individual, by participating in discussions, is influencing what kind of information becomes meaningful in the group. Thus, equality in discussions is an essential mode of working in group processing of individual knowledge construction.

Because a self-directing group does not have a supervisor to control its entire activity, the group has to do it by itself. The process of information processing by a group of individuals previously unknown to each other is a unique situation and its implementation will not succeed according to any automatic model of action. Maintaining and servicing of the group process requires submitting the modes of action to conscious scrutiny. In discussion, the activity should be looked at, the current situation assessed and future activities planned from the viewpoint of reaching the set goals. The use of the web forum, which was such a process of learning a new kind of action, should be critically assessed. Being used to authoritarian leadership in individual-centred information processing, group members are indeed often "happy" to hand over this task to a person responsible for it. Maintaining and attending to group process are tasks for which basic skills should already be provided during all forms of education.

In the information society the school system is faced with the great challenge of systematically implementing action learning. In education, everyone should be able, as socially equal persons, to develop their own individual knowledge construction, and extend that further. In the process, everyone should learn those basic information processing skills for networking in working life in information society.

Working life is in turn faced with the challenge of developing such modes of information processing in which the central significance of community is recognized. IT implementation applications seem to screen out organizing that creates community action from network-like information processing. Shared action by a group of human beings, such as shared knowledge processing, requires organization that maintains it. A challenge to research is to model and plan, more carefully than before, group-centred knowledge processing as a social phenomenon. When IT is itself applied to this knowledge processing, with its particular social and group nature, it may better serve knowledge-intensive co-operation in working life in the future.

Part III

Working and living in spaces

5 The dynamics of control and commitment in IT firms

Sirpa Kolehmainen

Transformations into the information society has been characterized by the importance of knowledge as a main factor of production and by the role of information and communication technology (ICT) as a medium of economic growth. Along with the decrease in the relative share of industrial production, the emphasis has been on services, especially on knowledge- and technology-intensive services, and on information occupations as sources of employment growth. Furthermore, a new organizational logic behind the competitiveness of business in the global information society emphasizes value chains based on inter-organizational networking in the innovation process. These trends affect work and its organization by changing the knowledge and skills base of jobs and occupations, and the means and practices of working as well as the employment and control relations within organizations, for example. Related concepts such as knowledge work, knowledge-intensive organizations or firms and knowledge-intensive business services have emerged to describe the changed content and organization of work within the information society (Alvesson 1993, 2000; Blackler 1995; Hautamäki 1996; Spender 1996; Karvonen 2000a; Rantanen 2000).

This chapter contributes to the discussion on information society by describing work and its organization in newly developed and rapidly growing innovative technology-based business services in the Finnish information industry. The focus of the chapter lies on forms of control within an environmental space of which increasing complexity, unpredictability and constant change are typical. Project-based client-oriented work in the area of the new information and communication technology challenges the traditional conceptions of control and co-ordination, especially those of manager–worker control relations. Knowledge work as complex problem-solving within projects characterized by changing commissions, tasks, working communities and even by changing physical places of work is not possible within traditional bureaucratic organizations with direct management control. Instead, it requires a creation of organizational spaces that supports both autonomous and shared knowledge-creation and learning as well as strengthening the organizational commitment of employees.

The issue of organizational commitment is an important condition for performance and efficiency as well as for competitiveness and innovativeness of organizations, especially in knowledge-intensive business services. Employees as primary generators of new knowledge are the most significant but scarce resource of firms. There still seems to be a lack of a competent and experienced labour force in these services. For employees there are open labour and wage markets that enable them to choose the best options. For firms, instead, there exists a continuous risk of human capital and knowledge leaving the organization. In the worst case scenario, leaving experts may take clients with them into their new companies. This chapter thus deals with the following questions: How does the client orientation of services formulate the work and working in high-tech information technology (IT) service firms? How independent and autonomous actually is the work in these firms? What are the forms of control in these firms – is the control strategy actually changing into a commitment strategy?

The analysis of work and its organization in high-tech IT service firms is based on six qualitative organizational case studies (Kolehmainen 2001). The data were collected mainly by conducting semi-structured interviews with information systems (IS) experts and their managers. The specific case organizations were selected within the software industry, new media and IT services complemented by a plastic product development and design firm. The cases represent independent private firms, whose main markets lie in the telecommunications sector. These firms can be classified into the category of technology-based knowledge-intensive business services (T-KIBS).

The space of the work organization is determined by a particular category of knowledge work. This is IS expert work which represents professional expertise in defining, designing, analysing, implementing and maintaining technical systems, and involves system designers, software designers, software specialists, application developers, graphics, graphic designers, designers and project managers, for example. The data consist of twenty-one interviews, of which thirteen were conducted with executives or people in other managerial positions and eight with expert employees. Expert interviews were complemented by a short questionnaire for all experts in the smaller firms and for experts in some specific ongoing projects in the larger firms. Of these, forty-seven experts responded. This study was carried out in 2000.

Features of knowledge intensity at work

In the information society, knowledge has become a main factor of production and it is supposed to affect work progressively in two ways. First, it upgrades the knowledge content of existing work within every sector of the economy, and second, it creates and expands new work in the knowledge sector so that knowledge workers and occupations come to predominate in the economy. In both cases, informatization of work, together with the development of information and communication technology, creates new ways of working and organizing work, thus changing work cultures as well (Kumar 1995; Castells 1996).

Special attention has been paid to a significantly growing part of the service sector which produces high value, namely the knowledge-intensive business services (KIBS). Many services are knowledge intensive in such indicators as the large number of professional and technical personnel and high level of investment in the new information and communication technology. Besides that, knowledge intensity of KIBS reflects integration of service-specific science and technology base and their contribution to knowledge formation in the economy more generally by producing and transferring new technology-based knowledge. The services that KIBS firms render are knowledge intensive, both in terms of complex problem-solving needed in the process of service production and in terms of required capabilities to interpret and use these services. KIBS firms mainly supply intermediated inputs into their clients' knowledge creation and production processes rather than outputs for final consumption. Thus, KIBS are growing as a part of the new logic of organizing production as networks of value chains between and among organizations (Miles *et al.* 1995; Hautamäki 1996; Haukness 1998; Karvonen 2000a).

What distinguishes KIBS organizations from professional organizations is particularly that KIBS aims at profit-making – their main clients are other businesses, including the private and public sectors and the self-employed – and that their services and knowledge are usually co-produced with clients. Examples of more traditional KIBS that are intensive users of new technology include accounting and bookkeeping, management consultancy, legal services and design. Computer and related IT-service industries can be considered as a main component in the category of technology-based KIBS, which are often highly innovative in their own right as well as facilitating innovation in other economic sectors. (Miles *et al.* 1995; Haukness 1998; Kautonen *et al.* 1998).

The concept of knowledge with its explicit–tacit and individual–collective dimensions, for example, is very abstract, broad and difficult to define (Nonaka and Takeuchi 1995; Spender 1996). Frank Blackler (1995) argues that knowledge should be understood as an active process of knowing that is pragmatic, situated, contextual and relational rather than regarded as something that people just have. Thus, while defining the concepts of knowledge work and knowledge workers, these should not be specified in terms of what the workers know but in terms of what they do in practice. Likewise, knowledge organizations should be defined not only in terms of their knowledge-based outputs and well-educated professional workers, but also in terms of how work is in practice organized so that it makes possible creative knowledge production and supports both the economic performance of the organization and the competence development of the employees.

All work and working requires knowledge and developed competence. Knowledge workers, however, form a special group which can be defined by its members as wage workers whose level of education is high, and who have access to knowledge that is formal, complex and abstract and the ability and opportunity to apply and process it. Their work combines both planning and implementation. It is related to creative identifying of novel problems and to producing unique, novel and value-adding recommendations. Through experience they have also developed contextual knowledge and competence needed in problem-solving. In addition to working on knowledge and information using ICT as the main medium of work, they also work on people during the work process (Reich 1992; Starbuck 1992; Scarbrough 1993; Winslow and Bramer 1994; Blackler 1995; Frenkel *et al.* 1999; Blom *et al.* 2001).

As owners of both explicit theoretical and tacit knowledge essential for the performance of the organization, knowledge workers exercise power within organizations. Knowledge work requires employment relations, a task structure and a co-ordination mechanism that allow for creative application, manipulation and extension of knowledge. Correspondingly, it can be argued that organizational structures and processes are influenced by the nature of knowledge and tasks to be performed within the organization. In essence, the less programmed and more uncertain complex knowledge and tasks are, the greater is the need for flexibility around the structuring relationships. An organizational form is also influenced by its environment. The higher the uncertainty and complexity in the environment, the greater the need for flexible structures and processes to deal with it (Tidd *et al.* 1997; McLoughlin 1999; Depickere 1999; Frenkel *et al.* 1999).

An alternative to traditional organizations with a mechanical division of labour and bureaucratic top-down control is a more dynamic, non-hierarchical and flexible organization that has mainly been characterized by the organization archetype of adhocracy developed by Henry Mintzberg (1989). The structure of adhocracy is fluid, organic and selectively centralized. Functional experts are deployed in multidisciplinary teams of staff to carry out innovative projects. Co-ordination is based on shared goals, mutual adjustment and expertise encouraged by the liaison personnel, integrating managers and the matrix structure. Furthermore, shared vision and leadership, creative climate, customer focus, inter-organizational co-operation, the roles played by key individuals, training and development of the staff, how the organization itself goes about learning and sharing knowledge and how it is involved in innovation are some examples of the factors which contribute to creating a more or less supporting organizational context. The environment of adhocracy is complex and dynamic, including high technology and temporary projects, and the form is common, especially in young industries such as the information sector. According to Tidd *et al.* (1997), the strengths of this project type of organization model are its ability to cope with high-level uncertainty, its creativity and its innovation capacity, whereas its weaknesses include the lack of control due to the absence of formal structures and standards.

Optimistic theories on work and organizations in the information society support the idea of convergence of the empowered non-hierarchical flexible network form of organization based on enabling technology and creative autonomous intelligence. In comparison, some theorists support a continuity view according to which changes in a work organization are variants of the regimented bureaucratic form. There are also some that argue that elements of work organization combine many different ways and forms, or even ways that are not yet predictable (e.g. Frenkel *et al.* 1999). Nevertheless, the features of knowledge intensity of work described above reinforce the impression of changing work organizations and control relations in knowledge work in particular.

Control and commitment as dimensions of work organization

The concept of work organization can be defined as a central forum that mediates and shapes up the content of work and working. Control of work as well as division of labour and different kinds of interrelationships among employees are considered to be central social dimensions of work organization. Organization of work concerns both intra-organizational and inter-organizational forms of action. In the service sector, clients in particular are an integral part of work organization, either due to simultaneous production and use of many personal services or due to a strong client-led definition and even co-production of the actual service in many business services. Work organization can be seen as a dynamic system operating within the workplace and comprising various elements that are related to the organization and society in which the work organization is embedded (Julkunen 1987; Frenkel *et al.* 1999; Ståhle and Grönroos 1999).

The analytical framework applied in this study is the ideal type of elements of work organization developed by Steve Frenkel *et al.* (1999) in their analysis of transformations of work organizations in the information society. Within the elements of work organization, the framework includes work relations, lateral relations and vertical relations. Lateral and vertical relations comprise the constituent setting of work relations, both shaping and being shaped by the features of work.

Work relations describe the typical features of work and the worker: the nature and complexity of work, the basis of work roles, and the competences needed. The key features of work are illustrated by clarifying what is typically worked on (materials, persons, information or knowledge) and by describing the nature of knowledge, skills and creativity required in the process of completing tasks.

Lateral relations deal with the functional or horizontal division of labour within an organization. First, they focus on relations between co-workers in the organization by describing the forms of working as well as the extent and nature of co-operation and lateral communication. These features illustrate the forms and intensity of a task and learning interdependency, integration of informal relations, and forms of teamwork. Second, lateral relations focus on the client interface, relationships between workers and clients that are often structured by management. The client interface emphasizes the client's role in a work process, especially worker–client–management interdependency from the standpoint of the changing nature of control in work, and its consequences for job satisfaction.

Vertical relations emphasize the hierarchical division of labour within an organization and refer to two aspects of hierarchical power: employment relations and control relations. Employment relations describe the conditions of employment, recruitment, working time, training, career structures and reward systems, for example. They are partly dependent on the national legislation, social customs and economic circumstances, and constitute a general form of control. Control relations in particular refer to the means by which managers exercise direct power over workers, on the one hand, and to the extent and patterns of worker participation in decision-making on the other.

Depending on the nature of work within the organization and the characteristics of the environment in which the organization is embedded, different forms of control prevail as part of organizational structures and processes. The existing forms of control in an organization are related first to management's knowledge of the labour process, and second, to management's ability to measure output. The role of clients especially in the service sector both as sources of information for control and as direct controllers is essential, and also has to be taken into account. Different forms of control also build up different social relationships and patterns of commitment. According to Etzioni (1961), coercive control creates alienation, whereas remunerative control brings about calculative commitment, and formal norms as a source of control create moral responsibility and commitment especially when the norms are internalized (Julkunen 1987; Frenkel *et al.* 1999).

Work commitment is a wide concept dealing with organizational behaviour. It includes a grouping of concepts that focus on various but interrelated aspects of commitment connected with work. These are: work ethic endorsement focusing on the importance of work itself; career/professional commitment emphasizing the importance of one's occupation; job involvement describing the degree of experienced engagement in work activity; and organizational commitment describing employee dedication to an organization (Morrow 1993). Organizational commitment is the most studied form of work commitment.

Three conceptualizations as components of organizational commitment are commonly used. Affective commitment refers to employees' emotional attachment to, identification with and involvement in the organization. These employees continue employment with the organization because they want to do so. Continuance or calculative commitment refers to an awareness of the cost of leaving the organization. These employees remain because they need to do so. Normative commitment describes a feeling of obligation to continue employment. These employees feel that morally they ought to remain with the organization. Also, other conceptualizations on commitment emphasize either specific rewards, identification with the organization's values, or internalization of the organization's values as dimensions of organizational commitment (Meyer and Allen 1997: 11–12; Depickere 1999).

Direct control includes formal behavioural control over those performing tasks, quality control, designation of authorization responsibilities, standard operating procedures, rules and budget, and expenditure guidelines. Direct behavioural control may be appropriate when desired behaviours and outcomes are easily defined. However, in cases of more complex and non-analysable tasks, output control is more appropriate than behavioural control. Output control has the potential for providing professional experts to exercise judgement with discretion and to be creative, but it also passes a great deal of responsibility on to the experts who might be reluctant to take all that responsibility. However, by developing group- and team-based work arrangements, organizations can encourage the development of the forms of social control derived from mutual commitments of group members to each other and to the shared ideas of the members. These forms of peer control are related to new management concepts such as empowerment, coaching and entrepreneurship. These principles require the building up of a strong commitment among experts to the organization by means of flexible work arrangements, for example (Adami 1999; Frenkel *et al.* 1999; Ezzy 2001).

Forms of indirect control include mechanisms described by employment relations, such as recruitment, socialization, career development and training as well as compensation and benefits. Forms of indirect control simultaneously represent the areas in which management can actively attempt to enhance employees' organizational commitment. Input control regulates mainly antecedent conditions of performance, ensuring that the employees' skills, knowledge, attitudes, values and interests match those of the employing organization. Along with the selection process, socialization of newcomers in the organization, training and career programmes are an important avenue for organizations to instil preferred behaviours and attitudes in participants. Challenging jobs with a high degree of autonomy are among the most important areas for organizational commitment, especially for knowledge workers. Further compensation and benefits can enhance the commitment and diminish the intention to leave the organization (Adami 1999; Depickere 1999).

The organizational commitment of IS experts within the turbulent area of knowledge-intensive technology-based services is not necessarily self-evident. One can ask whether they are committed rather to their technological competence and the hype of the new technology in general, instead of committing to an organization or even to an occupation or career.

Shared expertise in knowledge-intensive projects

The organization of business in high-tech case firms is mainly divided into two parts. One business area concentrates on software service production and consulting either as a part of their client's production developments or for integrating the client's business application. The other business area develops wireless software products for sale. These two main areas are usually subdivided into separate, relatively independent units by industries or functional service products, for example. Based on the business concept and internal division of labour, the case firms can be called (professional) expert organizations. These are flat hierarchies with a small number of managers and office staff, whereupon IS expert workers form the operating core and thus the majority of the personnel. There is high horizontal job specialization among IS experts, organized either formally or informally into specific competence areas needed in business. Behind the formal organization of business, the work of IS experts is group work or teamwork which is carried out within temporary projects based on client commissions. These case firms can thus also be referred to as project organizations.

Projects are not permanent organizations but based on changing client commissions. They have a clearly defined beginning, aim, implementation process and end. There are usually several separate projects going on simultaneously within a project organization. One project may also consist of several internal or external co-projects with at least some collaboration. In those cases with many partners in external co-operation especially, the client is usually the co-ordinator of the main project, and the co-projects with their own project management are implemented in distinct subcontracting companies. Clients are always an integral part of work organization within the case firms. The role of the client, however, varies between the projects from that of the buyer of the service to the evaluator of the outcome, or from that of the partner and co-producer of services to the co-ordinator of the whole project.

The subject of IS expert work in the case firms is software service products. Software as codes, languages, programs and networks also forms tools for performing work tasks, a means of communication, co-operation, documentation and management during the work process and also the product of ultimate work efforts. While the commission of the project in the case firms mainly focuses on novel problems, its implementation deals with uncertainty and complexity. IS expert work is thus characterized by theoretical knowledge about the new information and communication technology, with the need for contextual knowledge and experience due to the high-level client orientation of the services. A project group usually consists of experts from different competence areas and from different business units, depending on the demands of each commission. Project work indicates functional interdependency and requires both formal and informal interaction and mutual learning among IS experts and between the service supplier and clients as well. IS expert work is based on shared expertise and responsibility for attaining the goal (e.g. Lehtinen and Palonen 1997; Hakkarainen *et al.* 2000).

> We have big assignments which we divide into parts, such small parts that finally they are performed by only one worker. In that way, individual experts are important but the entity is composed only by putting these parts together ... An entity is more than a simple sum of its parts.
>
> (interview with project manager)

A project group can include experts from one organization only, but can also cross the organizational boundaries to clients and other contractors. Furthermore, when case firms have offices in different cities in Finland and also abroad, the project can extend to various locations as well, thus having virtual dimensions in its organization. The number of members within a project group varies from two to ten. A project consists of workers mainly with the same knowledge and status level. The duration of projects varies from a few months to a year, or even to several years in cases where service organizations participate in their clients' production development. Within longer-term additional development projects, it is quite usual for the composition of the project group to change during the project, because the experts want alternation and the employers want to keep the experts motivated. However, after finishing the commission, the project group eventually dissolves and the expert composition of the next project will usually be totally different.

An individual worker may participate in several projects at the same time. Within each project group, the position and task of the worker might, however, vary from project manager on one project to technical adviser on another, and actual project worker on the next project, for example. It is, however, more common to participate in only one project at once, because projects generally deal with complex, usually confidential problem-solving and new knowledge to meet their clients' needs. Instead of strictly divided job descriptions, IS experts respect more flexible and wide-scale competence. They typically seek their way to different jobs and positions within each project in order to get a challenging task, to learn something new, and thus to get wider experience in developing a new technology. Contextual knowledge and experience are equally important preconditions for IS experts' work. As projects may change quite fast, the combination of experts in each project and of their closest co-workers (at least on projects) may also change. In addition, depending on the commission and the clients, the place of work may be situated within one's own employer organization or within one's client organization, or the place of work can vary between the two according to the work process of the project.

In short, continuous change in work tasks and learning new project-specific knowledge, shared expertise alternating with more autonomous phases in problem-solving on project teams as well as changing expert composition and even changing working environment depending on each commission, all characterize the work in high-tech IT business services. These features of project work emphasize both spontaneous, communicative co-operation among IS experts and the need for formally organized co-ordination.

Feelings of autonomy within the system of control

Project-based knowledge-intensive work organizations challenge the traditional conceptions of control as well as of organizational commitment in many ways. The key relationships of work are no longer only those between the employer and the employee, but also those between the client and the contractor, and those among the members of the project teams in particular. On the one hand, knowledge-intensive expert work calls for autonomy and opportunities to influence one's own performance. On the other hand, the outcome of the work defined according to client commission is highly dependent on the work effort of the whole project group instead of that of a single worker. Work in knowledge-intensive business organizations requires close and continuous co-operation both with the clients and among the members of the project group. Further, experts do not take action because they are told to do so, to satisfy their superiors or for the benefit of a reward only. New motivators are goals, innovation and being part of creating something significant (Tapscott 1998; Frenkel *et al.* 1999; Alvesson 2000). IS experts within the high-tech case firms seem to be extremely committed to developing the knowledge of new ICT technology and living on top of the development of the technology.

> These workers are really technology buffs. All the time there's something new. Change and learning are somehow so typical of this work that those who don't bear it don't even apply for this kind of job.
>
> (interview with director of business development)

Work commitment is inevitably intertwined with patterns of control within knowledge-intensive business organizations. Control that is practised through a traditional hierarchical structure of an organization is not appropriate for expert workers. In an uncertain environment in which experts require autonomy and authority to decide about the most suitable methods and resources required to solve a problem, organizations should maximize their skills and functional flexibility by minimizing organizational control. This may mean flexible work arrangements and change from direct to indirect control methods for facilitating the skilled individual's task completion (Mintzberg 1989; Adami 1999; Depickere 1999).

There exist both direct and indirect forms of control in the case organizations. Within project organizations, forms of control and worker participation in the control process vary project by project, however. These fluctuations highlight the changing nature of commissions and the importance of clients for the working process. Due to the flat hierarchy and a quite loose division of labour by job descriptions, a lot of responsibility has been delegated to IS experts, especially to the project group and its manager. This promotes both mutual adjustment, "peer control" of the project group and self-control of an individual expert in the group because of common and shared goals on the project as well as socialized values and norms in the organization. In addition, the input control system, including recruitment, development and socialization processes, is also important in guaranteeing that experts have the capabilities and values to perform well and that they are especially suitable for the client-oriented project organization.

The main emphasis of management control in high-tech IT case firms seems to be on a result-oriented system after the project. Both direct behavioural control and output control before starting a project are problematic because each commission and each problem-solving process is unique in each project, due either to highly customized service products or to the development of new complex technological innovations. Occupational specialization of IS experts and the rapid development of technology narrows down the manager's knowledge about the labour process, thus diminishing the opportunities to programme the concrete working and outputs too much.

> Sometimes tools are ordered in advance and timetables are set up by the client for an already defined project. Thus, within these limits that are left, we have this one big entity which we can divide into parts and decide who will do each task. Not much is left for us when the client defines almost everything, specifications and documents … but we can comment on those and recommend alternatives.
>
> (interview with project manager/product developer)

Instead of management control, there might be a notable control from the client side. Especially in the definition phase at the beginning, the client may want to determine the specification of the whole process of the implementation of the commission, even to assign the experts performing the tasks. The role of the client and the formal forms of monitoring the implementation and schedule of an ongoing project vary depending of the nature of the commission and the knowledge base of the client. If the client does not participate in the project as a co-producer of the service, client control generally remains on the level of output evaluation. However, the role of the client is important because the pay eventually comes from the client. In many statements the client was actually perceived as an employer, because both the tasks and the pay come from there.

For each project, there is also a constituted advisory board with members from both the service provider and the client organization and, depending on the project, from other subcontractors. The aim of the board is to evaluate the progress of the project and to assess mainly the quality of results in each phase according to the planned goals, budget and timetables.

> The board assembles to meet once a month or every second month, and usually just checks on the situation of the project at the meeting. When everything goes well, the board meetings are just forums for discussion. But if something goes wrong on the project, it should be brought to the board meeting because it is the highest collective organ for problem-solving.
>
> (interview with director of telecom department)

The IS work process has been rationalized by formal methodologies and quality management procedures, which can be seen as forms of direct bureaucratic behavioural control. These structured methodologies prescribe the procedures and protocols to be followed in software development, the tools and techniques to be used at each stage, the timing and extent of user consultation at various points, and the required documentation (Beirne *et al.* 1998). As an integral part of their IT business, all case firms have to follow structured software development methodologies and processes. They also either apply (smaller firms) or have established the quality system (more established firms) based mainly on ISO 9001 certification.

Furthermore, depending on the stage of development, the case firms have formalized project management and working practices, or they are at least building up these processes. For example, a project can comprise such defined key phases as definition, design, implementation, testing and introduction, complemented by maintenance and further development. By using a structured project management which includes the intensifying of the delivery process on timetable, the managing of personnel resources, and the control of the cost-effectiveness of projects, for example, the organizations aim at ensuring the operability and quality of their solutions and at improving the efficiency of their business in general.

Rationalization of software development work has mainly been seen as constricting creativity and as deskilling IS expert work as well as being a means of controlling expert work (Beirne *et al.* 1998; Barret 2001). However, these formalized standards and processes also guarantee the documentation and reten-tion of knowledge in organizations if the experts leave. Furthermore, standards can also be seen to support and form a framework for conducting IS expert work.

> We try to create and develop our service processes. We try to create working practices that act as regular standards for experts at work. But new knowl-edge, however, arises in general from more informal and non-standardized innovations. Anyway, I'm a bit like a bureaucrat and I do support standards. I've always seen that standards and creativity are not the opposites. By developing and standardizing working processes and the quality of the service production we can only create the frame for working and business within such creative action as programming work still is. These processes must be the kind that don't limit creativity and innovativeness. Their signifi-cance is that the experts don't have to think about the details of their work every day, but they know what has to be done and can then release their creativity for the tasks that aren't specified. This is a young industry but the organizations grow quite fast. It just is necessary to organize the work and business but in a way that doesn't limit innovativeness.
>
> (interview with director of telecom department)

In the established high-tech firms, it is common that the organized quality system function, or at least the quality manager, formally supports and controls the technical and functional implementation as well as the quality of the service process and the output of projects. This function usually also includes in-house training of personnel for the working practices of the firm and for developing their technical and business skills. When working on new technology and tailored client solutions, output control in general is possible by means of a customer satisfaction survey after the project. If the project lasts several years, the client is usually asked to evaluate the project on a yearly basis.

Typically, the evaluation includes an assessment of the quality and economic efficiency of the work process on the project and the quality of the final service product as well. Sometimes, although the process has been successful, its results might nevertheless for a variety of reasons be unsuccessful. The evaluation might even include a performance appraisal, controlling the working and competence of each IS expert on the project. This assessment is connected to workers' wider career development but it might also have an effect on their rewards. The quality system is actually among the most important functions within business-to-business service organizations. The idea behind this pattern of control or co-ordination is to guarantee confidential and long-term client relations and partnerships, which for their part also motivate the personnel by means of continuous and challenging projects.

At the practical level, after being established, the project group itself and especially the project manager have the main responsibility for attaining the defined results. Project managers are the key persons on projects. They organize and co-ordinate the project, in respect to IS experts who work on the project and in respect to management, the quality system and the client. However, project managers have the main responsibility for the whole work process and for its continuing on schedule and according to plan, as well as for the final results. One of their main duties is to lead the expert group in a way that its members' competences are utilized. Project managers also communicate and interact in many directions, with the management of their own organization, the project group, the client and other possible partners. Project managers may direct and conduct several projects at the same time and their roles can change between projects. They can be just administrative managers or deeply involved in the technical implementation of project goals. The more the project manager is involved in the project, the better it is, because the manager may also be in the senior position, thus being able to give advice to other experts on the project. Among other things, how the project will succeed is highly determined by effective team leadership and effective conflict resolution mechanisms within the group, as well as by the organization of the work process and decision-making, which are the responsibilities of the project manager (Tidd *et al.* 1997).

> The project manager leads the whole project. But within it there are smaller teams that do things together, usually through team dynamics. If a newcomer does not know so well how to do the work, the others will then give advice and help because they know that the sooner the newcomer learns, the better the whole group will proceed with the work. These teams are so small that they usually control themselves.
>
> (interview with director of business consulting)

> Since everybody on the project works together for a shared goal, everybody also makes progress. Then if somebody starts to laze or something else happens, it is possible to react easily and say that you are now a little behind schedule. And to check how much behind and what can be done together

for it. We don't leave the worker alone but find out together how to get back on schedule. This kind of co-operation is actually very important in this job.

(interview with graphic designer)

While the project manager might be involved in several projects at the same time, the need for self-direction of the project group is evident and also a natural part of expert work. On the one hand, a project has phases for group work and shared expertise (e.g. defining, planning, testing). On the other hand, it also has phases for autonomous working (e.g. coding), in which live communication among team members is also needed and necessary. Formally, projects usually have weekly meetings in order to inform their team members about progress and possible problems, to follow up the budget and timetable of the project, and to discuss the programme for the coming week. The memos of these meetings are also sent to the client. Further, it is common in high-tech firms to establish a project-specific Intranet with limited entry for each project. In this Intranet, all project documents as well as each phase of the project are in full view and available for everyone involved. Also, through the Intranet the participants are able to discuss the progress and problems of the work. The project-specific Intranet enables first the co-ordination of knowledge and documentation of the process and the results of the project; second, it can be used as a means for management and control; and third, it is also a means for learning and development of know-how for IS experts.

IS experts conceive of themselves mainly as planners: that is, as adapters and revisers of innovations but also as operators performing tasks, either independently or with guidance. Despite the variety of control forms described, most IS experts regard their work as fairly independent: on the given project plan (in the planning of which they have usually taken part), they themselves can decide how to proceed in their work. There are no rules or direct supervision to control the concrete, everyday work of IS experts, who value highly the low hierarchy and flexible working arrangements and hours at the organizations for which they work. Instead, the most constricting factors in their work, according to IS experts, are project timetables. Contrary to the nature of knowledge-intensive business and expert work as client-based and collaborative service work, they also consider client-orientation, and especially the dependence on other experts regarding co-operation in completing tasks and achieving results, as constricting their own work. Limitations in flows of information – especially on the projects with participants not only from client organizations and other possible partners, but also within one's own organization in general – might affect the autonomy of an individual IS expert. Further, technical equipment as well as performance and quality appraisals were seen as constricting autonomy in everyday work.

Like knowledge workers more generally, IS experts have a considerable job and project-related autonomy on average, but they do not have much influence at a wider workplace or organizational level (Frenkel *et al.* 1999). They do have a certain clear influence on the content of their jobs, working practices, working pace and hours, as well as on their own career development and whether or not to participate in training. They also have some opportunities to influence the decisions concerning conflicts and order within projects, the content and extent of personnel training, and the wage and incentive system. However, they do not have much influence on management practices, on the definition of organizational goals, or on project evaluation methods. Furthermore, they do not have a lot of opportunities to participate in the selection of project managers or the composition of the project group.

The opportunities of IS experts to participate in decision-making and to influence their work varies project by project, depending on clients' activity. First, within firms' own production development, the significance of a client may be marginal and the designers only have to think about whether or not a client will buy a completed service product. Second, while participating in clients' production development in wireless applications, for example, specifications made by the client may be so accurate that they do not allow for much variation and room for innovations for the service provider. In these cases, the service product is unique and thus highly tailored to meet the client's needs. Third, providing Internet-based new media solutions for clients and solutions integrating the operative systems of clients, for example, the intensity of the clients' role and co-operation is dependent on their knowledge about the potentials of information and communication technology. There have been cases in which the client has given the service provider a free hand to specify and implement the service, only evaluating the final service product.

In any case, it is important that after the higher management of an organization has set up a project group, all experts can participate in the definition and decision processes in every phase of the project. The project building stage is a critical determinant for the success of the project. Besides clearly defined tasks and objectives, the commitment of participants to shared values and norms is governing the way the group will work and effectively perform their tasks (Tidd *et al.* 1997; Björkegren and Rapp 1999).

> The project manager and IS experts with the client representative define the project's goals together. This must be clear in the definition phase so that the designers can commit themselves to the project. It cannot be supervised by the project manager only, because he or she doesn't necessarily have such a clear impression of what the commission in practice actually demands.
>   (interview with director of business development)

Time is a contradictory factor related to control and autonomy in IS expert work. On the one hand, knowledge work has been said to be independent of time and place. On the other hand, projects are bound to more or less rigid timetables for everybody involved to follow. The most critical factor in project management is designing the timetable and then managing the project on schedule. On almost every project, timetables caused some problems; they were also seen as limiting the autonomy at work. Within the case organizations, the personnel's working hours were not controlled, and IS experts highly respected the flexibility of these working arrangements. However, the IS experts interviewed said that they do not prefer teleworking or working in the clients' workplace when given a choice. This is because in addition to project members they consider discussions with other colleagues in the organization as essential during the problem-solving process.

Moreover, IS experts said that they tend to work normal hours (37.5) per week on average, as agreed in collective labour agreements. Overtime is common in project work, however, especially close to deadlines at the end of a project. Thus their working time arrangements involve features of the culture of long hours typical of greedy organizations (Ezzy 2001). Usually their working hours per day vary between 9.00a.m. and 6.00p.m., during which the prime time in business in general is actually contained. Prime-time working is necessary for co-operation with clients and other possible partners, as well as for shared knowledge and co-operation between the IS experts in specific projects. The timetable and follow-up of working hours of each project worker is the basis for billing the client; they might also be the basis for rewarding individual workers, or alternatively the whole project group. Therefore, time is an important means for management in controlling the employees' working capacity, because there seems to be a higher risk involved in working overtime in knowledge work and within the IT sector in particular.

As control is necessary for effective functioning of business and implementation of work processes, it seems that some combination of direct output and indirect forms of control are present in most organizations. Knowledge-intensive organizations concentrating on producing novel and complex business-to-business services in particular are often representing a balance of output and indirect input controls (Adami 1999; Barret 2001).

While the personnel is the most important but most scarce resource for the high-tech IT case firms, the importance of recruiting appropriate employees is highly emphasized. The criteria for selecting new employees are quite demanding, yet varied and flexible. Technological knowledge is naturally the basic element. A formal degree is highly respected but not required. More important elements are personal attributes, engagement in new technology, social skills and the ability to learn.

> When interviewing new candidates for a job, I used to ask them what was the most important thing for them in work. Most candidates have usually answered that actually two things matter: first, success in work, that it is intrinsically rewarding; and second, the chance that it offers of opportunities to learn new things all the time. If candidates answer like that, they have good chances of being employed.
>
> (interview with director of telecommunications)

The IS experts in the case firms do have very special technical knowledge and competence with regard to ICT. However, they saw themselves as generalists who may have gained some special expertise through client projects. Their expertise develops especially through the needs of clients and practices in conducting their work with co-workers and clients. They also respect wide-scale knowledge because it allows for more opportunities to work on different projects and, at the same time, to learn what is new in the rapidly developing ICT. IS experts seemed to be very committed to their technical competence, in that they are able to maintain and develop further by means of client projects. Instead of extrinsic features of work, such as wages and promotion, they respect more intrinsic and social features of their work, such as challenging and changing work tasks and opportunities to learn, as well as the feeling of being on top of the development of new technology.

The president of the board of directors of one case firm attaches importance to understanding the reasons why and in what kind of circumstances workers are willing to leave their jobs.

> In my opinion, there are two things. First, IS experts are, on average, very eager to learn. If they feel that they don't to have enough challenges for learning, they change to another firm. Second is the fast pace of our business. In this area, occupations or work tasks usually change every second year because of the rapidly developing technology. Too similar and stable work tasks lead to the willingness to change employers. Thus learning and the changing nature of work are intertwined.

While project work in project organizations as such guarantees versatility and change at work, in the larger case firms in particular there were also formally organized intra-firm mobility programmes. Maintenance and development of knowledge of the personnel was recognized as a need and also established as part of business strategy within most case firms through internal and external training on technique, project management, customer service and business, for example. The case firms also emphasize individually targeted management of experts. The yearly or more frequent manager–worker career development consultations, for example, were considered an important tool to clarify how the experts really want to develop their careers in the firms. However, continuous and open communication between management and experts was seen as an important condition for organizational commitment of the experts.

According to Sveiby (1996), the business logic of knowledge-intensive organizations depends highly on how organizations regard their assets, how they attract their personnel and clients, and how they match this capacity for problem-solving with the needs of the clients. Walton (1980, 1985) emphasizes the interdependence between the organization structure and human involvement, according to which the innovativeness and efficiency of business operations are dependent on shared needs and goals between the personnel and the organization. Employees commit to their organization if it satisfies the work criteria that are important to them, such as involvement in decision-making, security, job challenges and autonomy, for example. Employers continue to maintain and develop an attractive organization only if it proves to be effective in economic terms. Thus, human resource management and personnel strategy are essential preconditions for creating loyalty and trust between the personnel and the company, and especially for the organizational commitment of the experts.

Concluding remarks

Every organization has to practise some form of co-ordination and control over work in order to ensure that its plans and goals will be achieved. Organizational change as a reaction to the development of post-modern information society emphasizes the shift from traditional bureaucratic organization, with authoritarian and coercive control of employees' behaviour, to empowered adhocracies, with normative forms of control manipulating the workplace culture and generating employees' commitment using norms and values as well as participatory organizational and job structures and practices as controlling mechanisms (Mintzberg 1989; Lincoln and Kallenberg 1990; Klein 1994; Adami 1999; Depickere 1999; Ezzy 2001).

Control relations, especially in knowledge-intensive client-based project organizations, are assumed to differ from the traditional top-down manager–worker conception of control. The environmental space of the information industry with high-tech business services is characterized by increasing complexity, unpredictability and continuously changing projects, in which the creative problem-solving required of professionals and special experts calls for autonomy with full opportunities to influence their own performance. Relationships among experts, clients and management can be characterized by reciprocal interdependence. Management has to rely on experts' problem-solving abilities, experts rely on management for resources and support, and clients rely on the quality of the service product at the end of the project. In project work, the key relationship is changing from the manager–worker relationship into the client–service provider relationship, and into interrelationships among the members of the project group, emphasizing the role of organizational commitment as modern control strategy (Frenkel *et al.* 1999; Depickere 1999; Barret 2001).

In a high-tech organization, quality management procedures, standardized methodologies of a work process and standardized project work practices can be considered as the most salient forms of direct (behavioural) control in IS experts' work. Standardized technological methodologies are actually so much an integral part of software development that supervision is based more on self-control than on management control. Quality control and control of project practices, instead, are supervised formally either by organized quality management functions, by a project board with members at least from the client organization and the service provider organization, and by the project manager. Furthermore, following up the budget and project schedules is a form of direct control. Depending on the project and knowledge as well as clients' activity, client control may exist both at the definition stage of a commission, during the service production while the client may evaluate the process and the quality of the service, and after the project when clients evaluate the quality of the output and process of attaining the set goal. Output control in advance is problematic, because of the ambivalence in working with rapidly developing new technology and because service products are not always very foreseeable. In addition, the work process in high-tech organizations producing business-to-business services requires close interaction with the client on almost every project, which also complicates defining the exact output, thus supporting the output control after the project.

In everyday work, the responsibility for organizing and performing tasks and the quality of work remains mainly with the project group and especially with the project manager. Project management and leadership is highly respected within the organizations whose business is based on projects. Thus, the success of each project and customer satisfaction stands or falls along with the ability and skills of the project manager to organize co-operation among project members. In addition to the forms of direct control and output control with client control, socio-normative peer control (Klein 1994; Frenkel *et al.* 1999) emerges from project work. Project work generates an internalized norm of reciprocity within a project group because it is based on shared responsibility and expertise. By establishing a project group with knowledge and mutual responsibility to implement a commission, management creates a structure in which the members of the group work together to meet organizational goals. This can be confirmed in advance by recruiting those experts who are willing to accept and commit to the values and norms of the organization as well as by applying human resource practices that support the socialization of these new experts in the organization. Moreover, in order to increase both affective and continuance commitment of experts, the organization can offer challenging jobs, training and career opportunities as well as proper compensation and benefits (Depickere 1999; Alvesson 2000; Ezzy 2001).

The assumption of unlimited time and place of work in the information society seems to fit IS experts' work only partly. Although they have flexible uncontrolled working hours, they have to follow project timetables, which almost without exception cause problems. However, instead of the project timetable, it is the experts who have to be flexible in their working hours. Nevertheless, IS experts tend to work normal weekly and daily hours despite project pressures. Co-operation in project work with co-workers, clients and other partners requires being at work prime time (from about 10.00a.m. to 4.00p.m.). Although it would be possible at some stages of the project, IS experts do not favour tele-working because they want to be close to their co-workers in order to get help easily in problem-solving. Their place of work may, however, change in different projects, while IS experts may work in their client organization, for example. Also, the project group may consist of experts from different locations and from abroad. This kind of virtual project work was considered somewhat difficult from the communication point of view. IS experts also emphasized the importance of face-to-face contacts in their work.

Nevertheless, at the level of an individual IS expert, these forms of control do not represent direct supervision; thus, they seem not to be constricting the experts' feelings of autonomy in work. IS experts referred to themselves as mainly planners who autonomously adapt to and revise knowledge and innovations, or as operators who independently perform tasks. Especially they respect the low hierarchy and flexible arrangements of work in their work organizations. Change and learning are their main motivations in their work. While IS experts, in general, have great jobs and project-related autonomy, they do not have much influence on the wider workplace level. What seems to be especially problematic is the communication and flow of information through the entire organization. However, IS experts are mainly highly committed to new and rapidly changing ICT and its development. Thereby they also identify themselves and are committed to the organizations that offer them opportunities to participate in challenging projects, for the implementation of which they are responsible from the beginning to the end, thus being able to learn and to be on the top of creating something new.

The focus of this chapter has been to describe the forms of control over the information systems (IS) expert work within high-tech IT business service firms specializing in software production of new ICT. The work of IS experts in high-tech case firms, as in most organizations in general, is characterized by a combination of direct or indirect input, behavioural and output control. Nevertheless, the emphasis is on input control, namely recruitment, training, career development and on client-led output-oriented systems after the project, as well as on the strategy of responsible autonomy in projects. The concept of responsible autonomy (Lincoln and Kallenberg 1990; Klein 1994) has been used to describe the efforts to commit workers to their organizations by ensuring non-economic rewards such as status, autonomy and responsibility within an organizational community. These forms of control describe decreasing direct supervision and increasing opportunities for employees to influence their own work, the goals of projects and the wider organizational level. They can also be seen as eliciting employees' organizational commitment, which has more and more importance because of its implications for the stability versus turnover and productive behaviour of existing employees as well as the recruitment of competent new experts.

6 Virtualizing the office

Micro-level impacts and driving forces of increased ICT use

Pernilla Gripenberg

Introduction

Organizations in developed countries are increasingly implementing advanced information and communication technology (ICT) and Internet solutions to support various organizational activities and increase organizational performance (e.g. Blom *et al.* 2001; Ricci 2000). Increasing ICT use is a process of institutionalization constrained by technology, its surrounding institutions and socio-historical context. It reaches out and both affects and is affected by macro-level change. While much discussion remains at a macro-level, increasing ICT use can also be seen as a micro-process of social interaction between individuals. In organizational contexts there are many different forms of settings to be studied. This chapter focuses on the office setting: how ICT and humans interact on a micro-level in offices and the consequences of this interaction for office work. The aim is to identify and describe micro-level impacts and driving forces. The ICTs of interest here are the most common office equipment, like the networked PC with e-mail, Internet and office software. The chapter illustrates aspects of these micro-level processes through interview and case material. The study is part of a larger study on ICT use in various contexts of everyday life (Gripenberg 2002; Gripenberg *et al.* 2003).

Throughout, I use the terms "virtualization" and "virtualizing" to refer to the continuous technologization process, going from ICT development to its use and re-interpretation in a specific socio-technical context. I distinguish between what could be termed the *virtual* and what is a process of *virtualization* that, through human choice of extensive ICT use, has a *potential to make the virtual*. "Virtualization" thus refers to the micro-level processes leading to *changes* in social contexts occurring as *consequences* of increased ICT *use* and *interpretation*. To the extent that ICT use and interpretation change the surrounding social context, these activities can be called virtualizing activities, i.e. interpretations, interactions and negotiations that support ICT use.

From my own perspective there is no "virtual" environment in the sense that there would be a *technologically determined* environment that could even start prescribing behaviour (Turoff 1997). Instead, it is through human agency, through the development, use and interpretation of ICT that the "virtual environment" is shaped, thus making it in a sense very "real" and very close to practice. A virtual environment is an environment in the "making"; just like any other environment or part of "reality", an organization is emergent and constructed socially through human interpretation, interaction and negotiation (see e.g. Ngwenyama 1998). Virtualization is no longer really about ICT use, but about new ways of acting and knowing in the increasingly ICT-supported environment, of which we still have very little understanding.

ICT use in organizations

The economic/rational perspective which assumes that ICT will increase performance and efficiency of the employees or the organization as a whole has not been very successful in explaining increasing ICT use in organizations. In some research the amount of a company's IT investment has, for instance, been shown to have a negative relationship to organizational performance (Byrd and Marshall 1997). In fact, any direct link between information technology implementation and organizational performance has been hard to prove (e.g. Gasser 1986; Kraut *et al.* 1989; Orlikowski 1993; Markus and Benjamin 1997) even to the extent that there seems to be no link at all. Rather, there is a need for completely different ways of thinking about both technology and the relationship between technology, organizations and change (Bjørn-Andersen 1988; Orlikowski 1992; Sauer and Yetton 1997; Truex *et al.* 1999; Atkins and Dawson 2001).

More humanistic approaches that assume increasing use of ICT in the organization has a positive effect on the quality of working life for the employees concerned – i.e. the "empowerment" argument – have also not been unambiguously successful in explaining the phenomenon. For example, as the technology alters the design of jobs, work processes, organizational structures, corporate cultures and individual skills, the result of information technology implementation into organizations often amounts to a whole range of unintended, unexpected, sometimes counterproductive, even paradoxical effects (Kraut *et al.* 1989; Orlikowski 1992; Robey 1997). In some instances, the effects may even be negative to the extent that the quality of working life becomes reduced along several dimensions (Tolsby 2000; Blom *et al.* 2001).

Another research stream on ICT-supported contexts and changing work environments is that on "virtual environments", "virtual communities", "virtual organizations" and "virtual teams". A recent review of this literature (Watson-Manheim *et al.* 2002) confirmed that this research is mainly conceptual, and that there is still a real lack of empirical articles, especially on more precise phenomena in work situations. In other words, in this stream there is also a lack of understanding of what actually occurs on the micro-level when ICT is brought into and increasingly used to support communication, knowledge sharing and work performance in organizations.

Yet another stream of research on ICT use and its micro-relationship with humans is the growing body of technology acceptance/adoption literature. In this, IT use in the organizational setting is seen as determined by variables such as users' perceived ease of use, or "user friendliness", of an IT system, perceived usefulness of technology (e.g. Hendrickson and Collins 1996) and personal characteristics such as "computer playfulness" (Webster and Martocchio 1992) and "flow" (Agarwal *et al.* 1997).

Much information systems (IS)/IT research is based on the idea of a perfect information system where humans are the only unpredictable factor and therefore should be reduced to a minimum number. Notions like "end user" (e.g. Rockart and Flannery 1983), "human factor" (Bjørn-Andersen 1988), "user resistance" to computer systems (e.g. Hirschheim and Newman 1988), "technology adopters" (e.g. Griffith and Northcraft 1993; Lassila and Brancheau 1999) have set the tone for how individual users are often seen as the passive "receivers" of technology in this research.

What is lacking in much previous research is a holistic view on the situation of individual users, where they would be included on their own terms and technology would be viewed as something to fulfil their needs, for example, contributing to the "human endeavour" (Bjørn-Andersen 1988; Norman 1993) or recognizing users as a source of IT innovation (Nambisan *et al.* 1999). Given that the reasons for increased ICT implementation and use in organizations are not clear-cut and that the social effects may be very broad, continuing long after implementation and initial use, it is surprising how little empirical research has been directed towards identifying micro-level driving forces and longer-term impacts of ICT in organizations.

Some exceptions indeed need to be mentioned. Zuboff's (1988) classic book on the computerization of work tackles marvellously the complexity of how IT is altering the nature of work and the dynamics of the workplace and how computer technology is fundamentally changing the nature of work and power. Burkhardt has studied longitudinally the role of interpersonal relationships in spreading beliefs, attitudes and behaviours in an organization following technological change (1994) and the effects of a change in technology on social network structure and power in an organization (Burkhardt and Brass 1990). According to Burkhardt and Brass (1990), early adopters of the new technology were able to reduce uncertainty for others, and this ability enabled them to gain centrality and, through that, power in organizational networks. They also found support for the view that diffusion of the technology occurred as a *result* of the restructuring process, in that individuals adjusted their patterns of interaction in order to learn from those who had already learnt how to use the technology. They suggest that competence-destroying new technology is less likely to be adopted early, at both individual and organizational levels, than competence-enhancing technological discontinuities. They conclude that the process of change seems to be the same across studies and that change results in various forms of uncertainty that need to be reduced individually and organizationally. Accordingly, this chapter takes a dynamic rather than passive view on the humans who are affected by increased ICT implementation, and also tries to see how these humans by their own actions sustain and increase this use.

Methods

The data collection was exploratory and inductive in nature, taking the increasing use of ICT in the office setting as the phenomenon under study. Two main methods were used: an interactive researcher-practitioner workshop, with six ICT users, and a case study, including three ICT professionals and five ICT users, which followed up on ideas generated at the workshop. The main context of data gathering was the business school where I work, which is both a traditionally information-intense and (increasingly) an ICT-intense organizational context. These data address how office workers experience the effects of increasing ICT use on their work and organizations.

The workshop was part of the activities of my department for collaborating with companies. Collaboration partners were invited to a workshop with the subject: "What does your advanced information technology really do to your company?" The workshop comprised a CEO of a small ICT company and a manager in a large ICT company, and four researchers. All were using ICT extensively to support their day-to-day activities. The aim was to discuss three questions:

- What is advanced/interactive information technology?
- What happens in a traditional organization when this kind of technology is brought in?
- How does that change the way we view the organization and how we should manage it?

The discussion evolved closely around micro-level, personal experiences of ICT use and how it affected, in an overwhelming way, a variety of aspects of office working. Thus, what really was being discussed was "What does ICT really do to you and to your way of working?" Drawing from the findings in the workshop material, an interview guide for a second study was made. This case study focused on the increasing implementation and use of ICT in the business school and the work of the interviewees, and the consequences of this on their perceptions of what they were doing, how and why. The timing was triggered by an upgrading of the main server, which caused some major system disruptions and breakdowns over a period of a couple of months. These breakdowns provided an excellent opportunity for interviewees to reflect not only on their daily use of ICT and its effects, but also on the institutionalization of this use.

Because micro-level ICT use and interpretation reflects its surrounding context, it seemed important to get as broad a picture of the case as possible before conducting the actual interviews with users, in particular about the ICT situation in the organization. Background information about ICT implementation and use in the business school was gathered through interviews with the director of the computing centre and other employees. Five further people were contacted and they agreed to be interviewed: a department secretary, an assistant professor, a librarian, a research director and a project manager. The interviews

were semi-structured around themes (Table 6.1), and the interviewees were
allowed, and sometimes encouraged, to drift beyond the original questions, as
long as the discussion stayed within the subject of using ICT. Each interview
lasted around one hour. Interviews were recorded and transcribed. In addition,
because I was working at the business school and had also been involved in
different ICT development and implementation projects, participant observations
and informal conversations were also a natural, additional part of data gathering.

The analysis of both workshop and case material evolved around some major
themes:

1   *Identifying actors and critical incidents* that had had an impact on or had
    somehow changed the use of ICT, either from the perspective of the indi-
    vidual or more broadly in the whole organization.
2   *Relations between actors, entities and objects* (e.g. employee–computer,
    employee–peer, computer–work task).
3   *Expressions of emotions and changes in relation to ICT* (directly or indirectly),
    focusing not only on specifically expressed emotions in regard to or in uses
    of ICT, but on what could be perceived from the text as describing or
    containing an emotion (e.g. expressions of frustration, excitement, enthu-
    siasm, fear) or a change of some kind (something moves, happens or
    changes).

Table 6.1: Interview themes

Theme	Type of information requested
Background information	Information about position, time of employment, tasks, etc.
Understanding of task and functions	How had ICT or "virtualization" affected one's own work, or alternatively, the school as a whole?
Understanding of position in formal/informal role system	How had technology changed one's own role at work, or in one's department, working community or the school as a whole?
Embedded infrastructures suddenly visible	How had the major disruptions, caused by the installation of a new server at the school, affected one's work and perception of technology?
Virtual organization	Was there a virtual school and what was it like?
Driving forces behind ICT use	What had been major events or who had been central figures that had advanced the use of ICT in the school or department, and why?
Attitude towards technology	What kind of relationship did the interviewee have with his/her computer?

Some macro-level forces have had extensive effects on the micro-level use of ICT and therefore deserve mention before going into the micro-level analysis. For example, there was a national decision that all Finnish libraries, including school and university libraries, would be connected in an electronic network. This seemed to be a major step towards virtualization. Another critical incident was the Ministry of Education decision to renovate the business school, so that existing systems could be expanded and new cables installed for more bandwidth connecting to national and international networks. Other forces of change came externally through collaboration. For example, the research director noted that the Ministry of Education had started requiring certain information in electronic format, no longer accepting paper copies.

Another issue in the business school was the difficulty in attracting competent people to the IT department. Because the school is state owned, the salary level is not very competitive compared to market salaries. The large turnover in personnel and the low experience level in turn made it difficult for the IT department to focus on developing long-term plans and solutions. Providing sufficient user support also suffered. Comparable but different external forces could be seen among the workshop participants, such as national and international standards directed towards the development of products and the fierce business competition putting great pressure on continuous change and development.

Micro-level forces driving ICT use

This section identifies micro-level actors and processes driving increased ICT use, and illustrates how interaction between individuals influences the use of ICT across the organization.

Learning in practice

The interviews suggested that both using and learning to make better use of information and communication technology, and thus its increasing use, was a truly iterative process at both individual and organizational levels. Change, or learning, at the individual level really occurred *in practice* – in the specific situations where work was done and where people interacted with each other. It seemed that, driving the learning, there was either a problem or a frustrating situation to solve, or there was a sense of victory on doing something in a new and "better" way than before. One interviewee (a woman in her fifties), spoke about how to learn new things to do with the computer:

> It is in practice you need to learn. I think that really the best way has been … many things … but that you sort of show … or anybody [strong emphasis] can come and tell that "this thing you do like this and this" and that is really a lot easier than if you sit in a computer lab and there are a number of people doing a number of different things for three to four

hours, and when you get out of there, then already the next day you have forgotten almost everything and are wondering how was it again.

The research director told a story about a colleague:

Take, for example, Professor X, who only a couple of years ago didn't know how to save a file on the hard disk, and never had been doing this, who then got a collaboration partner from another country who sent e-mail and files and this and that, and this resulted in that, today, this person even reads e-mail.

One interviewee (a woman in her fifties) stated with emphasis:

The more you learn, the more *fun* it is to use it [the computer] – that's really how it is!

and, regarding the Internet,

But I think really that the more you – one – learns, the more *positive* it becomes. And one perhaps becomes more interested, to a certain degree.

Another interviewee (a woman in her forties) commented:

If you use your computer as a typewriter and possibly as a mailing machine, then why would you start spending time on updating some web pages? But if you notice yourself how irritating it is when you go somewhere [to a web page] and you don't find that information and it is not updated and how *glad* you become when you find something that you notice can save you several days of work just because you happened to stumble over that page, then you start realizing that it can be a good idea to put some information about things up yourself on the web.

The role of local ICT enthusiasts

Internal pioneers (the human agents or early adopters) seemed to hold a key role in this iterative, social learning process. These people were not necessarily formal technical experts but technology enthusiasts: ordinary people working in the local context, but with a heightened interest in technology and technical gadgets compared to others in that context. Enthusiasts seemed to have the ability to translate the opportunities provided by information technology into an opportunity to improve the everyday work of their co-workers and other local users. As Burkhardt and Brass (1990) note, they decreased the uncertainty that new and complex technology brought about. These enthusiasts seemed to take on the role of teachers and advisers when teaching and advice on technology was needed, and seemed to be doing it simply out of their own interest.

A positive force of getting pleasure, both out of teaching and giving advice and out of learning from these teachers, seemed to be driving the learning process. Almost all the interviewees in the business school case and the practitioners in the workshops had at least one story about how some other person (a colleague or someone at work, a friend, relative or family member) was interested or knew this and that about a new program, a feature, a "trick" in a program, etc., and how this person passed on their knowledge to anyone who wanted to learn or was interested in listening. This perhaps tells something of the surrounding culture: that it is common to talk about technology and the interaction with it, and that being knowledgeable about ICT is something to be proud of and admired by colleagues.

In their way of using ICT, one of the interviewee's departments could be seen as a pioneer department compared to others in the business school. It was among the first to provide information on the web with extensive use of links, and had been the first to use advanced mailing lists etc. The following interviewee (a woman in her forties) told how ICT use became advanced at that department, and highlighted how development and learning is tied to social interaction between individuals and seems often to be quite coincidental.

> I can tell you exactly how it went – it's really simple, and it's really – and there you really hit the nail – that there is always, or you always need a kind of pioneer ... and then it is also a coincidence ... Well, you know X [a researcher at the department], she has always been interested in new things with the computer, and she just got interested in this thing, and she was probably ... yes, probably one of the first persons at Hanken [the case study organization] to start making web pages. There was probably, like, Y [former employee at the computing centre] and her who made these things. And then I realized, when I was watching from behind her back and because I have been doing a lot with computers, that it wasn't that difficult at all. And then when it was time to start making them too and we sat there with only a thin wall in between us, so then she came and showed me, and if I sometimes got lost I just shouted, 'X!' and then she came and showed me how I was supposed to make an image or something like that, and then it started working out. And then – now I in turn have started to spread this knowledge forward here. You know, I don't believe in this, that you should go on a course – or it's good to go on a course, but something like that, all personnel should be gathered and then someone would start *showing* how you do things in practice ... but then I would not have started doing this if I hadn't realized that, for example, all European information about Z [specific topic relevant to the position of the interviewee] was found in web format, so why would one then start doing it with pen and paper here at Hanken? So I had easy access to a teacher who quickly taught me this, plus I realized that elsewhere this was done like this: virtually.

Also the next story, telling of a contrasting situation, highlights the same thing: one interviewee (a man in his thirties) felt that his department was lagging behind in having courses, course material, slide shows or other things on the web or other computer- supported teaching:

Interviewer: What do you think is the reason for this, other than that you are afraid of it not working when it needs to?

Interviewee: The reason at the moment is that, or I would say, it is because there is no one at the institution who is knowledgeable enough. That someone would take this as his or her thing, to drive the institution forward as a whole: someone who would say that "Now we are going to have all our courses on the web, and I know how it's done and it is not difficult at all" ... But there is a lack in knowledge and interest to get this changed or to learn this knowledge yourself in order to get it done. One has other things to do, and as long as the old system works you just go on with that. So someone who would kick this off would really be needed, and then it would get going.

As Burkhardt and Brass (1990) note, people who saw an opportunity in the new technology and who could easily decrease uncertainty and make use of the technology in a competence-enhancing way, on both individual and departmental levels, became the early adopters. They were the ones who made the knowledge trickle throughout their departments, or to people who were physically close by or with whom they were interacting otherwise, because of work or some other relationship.

ICT knowledge as a source of power to influence decisions

In the case study, regardless of the formal authority, expertise or assumed critical knowledge about the technology that enthusiasts held, they seemed to play a surprisingly important role as influencing forces behind technological advancement at individual, departmental and organizational levels. The capacity to act as a translator between technology users and those who develop, implement or make decisions regarding technology in the organization is a new source of organizational power, in terms of influencing decision-making at the individual level (Burkhardt and Brass 1990).

Enthusiasts were by no means critical and objective, but there seemed always to be a degree of personal belief in the choices that these people advanced. However, these choices could coincidentally influence ICT use and development in the organization. For example, in the case of the pioneering department when it came to web pages, it seemed that one person, by chance, actually had a huge influence in setting up a norm, or best practice, for what departmental web pages should be and look like at the business school. This may not be an uncommon story on how important strategic decisions are made in organizations, especially when it comes to innovations, that no one really has an in-depth understanding of, as is often the case with new technology. Herein lie, of course, the greatest risks and challenges with the increasing technology use.

The risks of enthusiasts gaining too much power, with lack of real knowledge and insight into the type or number of consequences on the organization from the implementation of a new technology, or even from a new use of an application, can be high, and the effects can be critical. As one interviewee (a woman in her forties) explained,

> I was enthusiastic here a couple of years ago … why should Hanken print all these PhD theses … that it would be enough to print a small edition and then put the rest on the net. It's not difficult to make them in PDF-format and then just put them there. But now I have more and more started to understand this thing, and especially now that I am taking part in this European project, that there are very many of these intellectual property rights questions, and that I don't think we really have … or that it's again not just that you have a machine and a program and then you just clutter something there, but you have to consider all these things and make sure it is not possible to copy, to take or to change, and consider who has these rights.

Quite clearly, however, in the business school case, the power of influencing decisions regarding implementation and use has shifted away from the ordinary user and the former early adopters towards the growing IT support department, established in 1993, which is now taking care of IT purchase, installation and service functions centrally. Major decisions, in turn, are now made by the information and technology advisory board, which was established as late as 1999 as a joint board, consisting of people from the library and the IT support department. Very recently some possibilities to influence decision-making have been given back to the users, but in a different form. During 2002 a five-year ICT strategy formulation process has been implemented, with several working groups preparing different parts of the strategy proposals. The working groups have consisted of people from various departments, and the proposals have then been circulated through all departments for consideration and comment. In this way the decision process has perhaps become more transparent and democratic, including a more critical examination and debate of the development of the ICT situation in the organization. This to some extent reflects how the importance of ICT is growing throughout the school, and also shows that the need for grounding decisions among users is regarded as an important part of the development process. However, considering the huge changes in the ICT situation from 1995 to 2000, most probably with quite unpredictable outcomes, it is clear that any strategies regarding ICT need to be revised at least on an annual basis.

ICT: the new "friend" …

Considering that some of the users had, in their own opinion, a very neutral and tool-like relationship to their computer, while others felt it almost to be a part of their body, using ICT in one's work seemed to evoke surprisingly strong emotions, both positive and negative. It might seem far-fetched that this would have anything to do with how technology is institutionalized, but I argue that it has. When institutions work, you either don't think of them at all or you are content that they work smoothly – they are part of the surrounding system. Sometimes you reflect on things like "How could people ever live without bank machines?" (Or washing machines or cars or planes or supermarkets, etc.) But when they don't work it tends to be extremely frustrating and irritating, precisely because we don't really know how to survive without them – they are institutionalized to the extent that we have forgotten about them, but have also forgotten how to act without them.

On the question of what kind of relationship one of the interviewees (a woman in her forties) has with her computer, she started hesitantly and then burst out in an emphatic statement that: "Well, it [the computer] is a part of my body and my identity." The statement was followed by laughter and she then continued, "I haven't yet transplanted it as a third arm but ...", and then, hesitatingly, trying to formulate what to say next, she stated: "There are two things out of these kind of appliances that I would stick to for the rest of my life, the washing machine and the computer – with Internet access, that is."

One interviewee (a woman in her fifties) describes her relationship as being a "love–hate relationship", while another (a man in his thirties) describes the relationship like this:

> I don't have it [the computer] with me in my bed, but ... it ... I see it ... as a tool, but still I am actually quite fascinated by it. However, I know so little about it that for me it remains a tool ... I don't really have any interest in what happens inside the jar – as long as it works I'm happy!

It seemed that the more the computer was used and worked for the users, the more the user became "friends" with it and dependent on it and its functionality, and the more it started to take on a life-like shape. This was not as much seen as a separate "individual", but more like a limb or a part of oneself, something without which one was not whole, something that one became dependent on and expected to be there to function normally. These quotes illustrate how Finns cherish their tools and technologies, and this fascination is certainly one of the driving forces of increased ICT use, not only in these organizations but in Western society at large and Finnish society in particular.

... or the new "enemy"?

As long as the ICT works it seems to be fascinating, but quite often it does not work. These are the moments when the increasing dependence upon technology becomes obvious. One interviewee (a man in his thirties) reflects on the breakdown due to the server change at the beginning of 2000, and his own reflection makes him aware of the incredible pace of development and increased dependence on IT from 1995 to 2000:

Then [in 1995] we didn't know what it [the WWW] was, we had to invite
someone here to tell us about it, and still we didn't really understand what it
really was about ... Five years ago we didn't know what e-mail and the
Internet is and now, if you are even one or two days without it then everything stands still! ... You even start wondering if you should go to work at
all, because nothing goes forward [with breakdowns]!

Another interviewee (a woman in her forties) expresses her frustration with
the malfunctioning network quite strongly (emphasis added):

Today and yesterday there have been these *nerve-wracking* hours when I have
been thrown out from the net ... *you feel handicapped*. But of course it's a habit
that you get used to, and if you didn't have the computer you would find
alternative ways to do things.

Interviewer: Is it in some way scary to be dependent like that on one's computer?
Interviewee: No, I think that ... well, there are worse things to be addicted[1] to.
[Laughter.] No, it really makes things so much easier. But then we
need to be careful ...

She goes on expressing some fear that people might stop communicating with
each other, because it is easier to send an e-mail to the room next door than
going over and talking. One interviewee (a woman in her fifties) describes the
huge frustration due to the breakdowns from the installation of the new server
through an example of a problem situation when an urgent document, sent as
an e-mail attachment from China, first couldn't be opened and then couldn't be
printed because the network was down. When the printer finally worked, it
printed things differently from what was on the screen:

I couldn't get it completed on Monday, and then yesterday, when half of the
day had passed, I thought, "My Lord, one and a half days you sit here and
you can't get the help you need!"

Interestingly, the interviewee distances herself to some extent from the
computer and gives it life by letting it be the performer of the unwanted action,
when she does not understand why things are going wrong.

Users seemed to have no or very limited control over what is inside and
behind the computer – the cables and connections to networks of other
computers, printers, servers and other things that make the computer and its
applications function as expected. As illustrated above, this dependence makes it
difficult to be in control over one's own work and renders people helpless,
because they cannot solve problems without help from the IT department. If
things do not work smoothly there, it sometimes becomes really difficult to
perform one's ordinary work tasks. Simultaneously, as more ICT is being implemented and used, the demands on service and help from the IT-department

increase. The attitude may even develop that, since all this technology has been implemented and has to be used, someone else has to make sure that it also works properly. A man in his thirties explained:

> It is not my job to, so to say, get the tool to function, but there are experts for that. I mean, my time shouldn't be taken up by that, but if something goes wrong I only need to call the IT support and wait for them to get the problem fixed. [Followed by laughter indicating that this is how it should be, but that it is not always so.]

This may reflect an attitude against having to understand all the details about technology, meaning that it should be enough to learn how to use the tools. As learning new things requires a lot of individual and organizational effort, this kind of attitude seems reasonable, but it does leave the individual more helpless and dependent on others to perform their job with required technology.

In the instances above ICT had become institutionalized and, in a sense, to some extent invisible (Star 1999) as long as it worked all right. Even then it was somehow described as something one could depend upon; it was even fascinating at times. Yet this dependence was made visible by people not knowing how to perform their tasks without the technology, as when the technology malfunctioned. Then especially, users seemed to have a need to make sense of (cf. Weick 1995) and explain their view on technology, for example by giving the technology a life of its own or saying that there were experts to take care of problems. The examples also illustrate how, with more and more complex technology, individual users have less and less control over their work and the tools they are using to perform their work in the virtualizing organization. Because of this there is no way that everyone can understand and learn everything; people tend to stick to the very basic functions needed to be able to perform their work.

Micro-level impact of increasing ICT use

This section focuses on common themes that came up in the interviews on consequences of the virtualization process for the way work was conducted, and also tries to identify activities that could be seen as virtualizing activities.

The socio-technical context and its historical development

Increasingly, implementing and using ICT in a traditional organization may result in conflicts between the new and the old, and in confusion regarding almost every aspect of workplace activities. From the interviews it became evident that in the newer positions, without existing systems of tasks and routines, built more or less from scratch, it was easier to take advantage of new technology in a fresh way. In these positions, or with new people starting in old positions, the shift towards increased use of ICT was easier and seen as a positive challenge, whereas old systems that worked satisfactorily were harder to change. A new technology in an old position can be seen as a competence-destroying discontinuity, whereas in a new position or with a new person there is not so much old competence that could be destroyed (Burkhardt and Brass 1990).

The research director reflected:

> When I started working here with my new job, it was just when all these new things were happening and we started building up this infrastructure at Hanken – for example, information about research funding we only [strong emphasis] gave in electronic form and forgot about anything else … Every time I started on something new I started only in this new way. It would be different if I would have first been doing this job for, say, five or ten or fifteen years, because then I would have done things differently.

In fact, where new systems were implemented on top of old ones, they created ever more confusion, and at some places parallel systems were kept to support each other. The overlapping of old and new systems seemed to result in much uncertainty and in the duplication of both work routines and information, but also in who was responsible for new tasks and for teaching/learning them.

This situation not only reflects how history-dependent technology use was in the business school, but also says something of the freedom in the academic world and this business school in particular. This kind of freedom and high uncertainty tolerance, without management acting to control, may be different from business organizations, where personnel may be required, for example, to do mandatory training on the new systems. On the other hand, the whole case reflects something of the newness of the situation in both the business school and more generally. Management did not perhaps realize that there is a need for a person to be in charge of the web, the Internet or ICT-related tasks, including training and learning how these worked, because this was not common when these kinds of systems were first implemented.

The instability of new technical and socio-technical systems also seemed to be a source for part of the insecurity and duplication. As the main server kept crashing (also one of the reasons that the old server was updated) and files kept being destroyed by viruses or sloppy work or even lack of knowledge of how to store files properly, it seemed quite natural to keep paper copies of every document, e-mail message and other files, just in case. As an interviewee (a man in his thirties) noted briskly,

> Well, I have to say I find it hard to think about the day I would read everything on my computer. And I also think it is a question of security, it is like … I mean, hardware crashes and things like that, so there is always a security in having it on paper. All that can happen to a piece of paper is that you have water damage, or that the house burns down, but that really happens considerably more seldom than that your hard disk crashes!

Security and stability are, of course, also a question of learning and developing new understanding of how to store files and maintain servers and hard disks, which did not seem to have occurred in the business school yet. This also reflects how technology was easily implemented without considering training issues or security issues in advance, nor the long-term (or short-term, for that matter) stability of the system, including both its technical and social sides.

Information and paper overload, and difficulties of staying focused and prioritizing

Despite the notion that ICT can support the paperless office, in the case study the increased use of ICT seemed to result in more – or at least not in less – paper being printed and circulated. This was again a question of how things had been done before and how they could be done now with technology, and there seemed to be a lack of understanding of how to use the technology in order to support work instead of duplicating it and making it more difficult and messy. Instead of choosing one optimal medium for distributing a message, all possible mediums were now chosen. Also, the ease of getting and finding information that may at some point be needed seemed to generate a lot of printed paper that sooner or later ended up in the dustbin, even without being read.

As the assistant professor noted:

> Also this, that you can download an article and print it out, instead of standing there and photocopying it from a journal. It is really easier just to hit a button, but this has of course affected that one gladly … or I mean that this ease to gain access to things makes it so that one sometimes prints out things that one doesn't really need, and there it is, left cluttering your office.
>
> …this ease of printing makes you print out a paper, then you read the abstract, and then you throw the whole pile of paper away. Also, I think there is a lot of double information: it comes through the net and then on paper. In the worst case you print out the e-mail on paper and then you get the same information on paper in the internal mail, and in the end you throw all of it away.

Clearly, there was still a lack of information strategy and understanding about how things could be done more effectively with ICT and for what purposes it was best used. Which medium to choose for communication seemed to generate some confusion, both in the case organization and in the companies represented at the workshop. This seemed to result in an information flow that was sometimes difficult to handle, making it increasingly difficult to set priorities. For example, during the workshop there was discussion about how easily a blinking e-mail on the screen grabs people's attention away from whatever they are doing, fragmenting concentration and making people jump between tasks more than before. Because of all the information and communication channels grabbing one's attention, it seems to be more difficult to stay focused, thinking about what really is most important and in which order tasks should be done. The increasing amount of junk mail did not make the situation any easier.

It was not only those who receive huge amounts of information who found this problematic. Some of those who sent out a lot of information were concerned with and reflected upon this too, because they were aware that if they sent too much information of the type that the receivers were *not* interested in, they would stop reading the messages. For example, when mailing lists were first used, information, for example about a PhD course, could be received up to four times by one student. Some caution has since grown concerning the forwarding of messages initially received through mailing lists. Subscribing to too many mailing lists had proven to become a nuisance for some of the interviewees: when traffic grew on the lists they were soon overloaded with messages. This required the interviewees learning how to restrict the information flow in terms of both sending and receiving.

Several references were made on where information was supposed to be stored. The following extract from a discussion at the first workshop is a pressing example, both of the increasing amount of information and how to handle it, and of the never-ending task of ordering and structuring information in the organization. Yet the information does not seem to really make anybody any wiser:

Researcher A: Now, when you have told us about all your newsgroups and things like that, one can really see how all this information is flying about there, and there is so much of it. It's then only a question of grouping it the right way. Could one think about if people know more, then? Reasonably, they should know ... or ... no one can, you know, in this kind of environment, then say, "Why didn't anyone inform me?"

Participant (CEO company A):   Yes, I guess people know more ... they know [hesitantly], but they also get more frustrated over all the time it takes to look for, if you don't get it structured well enough.

Researcher C: We have to say that we [the researchers] belong to the great sceptics to this notion of "know", until you can say it with a little more emphasis, that "we know more".

Participant (CEO company A):   Yes, but I did shake in my shoes a little ... [General laughter in the group.] The amount of information that people have is larger than before, that I believe I can say without shaking in my shoes. But whether one ... like, the demand on the amount of information you need to know to act sensibly ... But the fact that you know how to handle the information doesn't help because the need for how much information one should be able to handle has grown even more.

In sum, the information structure as a whole, including which media to use and where to post messages, when to use e-mail, when to print paper copies, how to sort, file and store messages, in both e-mail and Intranets, really was an issue of concern and debate. This issue ranged across all levels, from individual decisions on how to sort and store e-mail messages, to department level, on how to structure departmental information and distribute it, to the organizational level, on how to build up company web pages, data bases, etc. These activities can perhaps be summarized as truly virtualizing activities, i.e. continuously interpreting, re-interpreting and negotiating how the existing ICT is supposed to be used and why.

Shifting and increasing (?) responsibilities

Interestingly, with the choice of electronic media it seemed to be possible to quite subtly shift responsibilities away from oneself and towards someone else. The reverse was, of course, also true: others shifted responsibilities over to one. One of the participants in the workshop noted that people had started to send copies of e-mail messages and even whole discussions to their boss. This generated a lot of unnecessary e-mail traffic for the boss, who certainly was not that interested in reading long e-mail discussions just because at some place in it there was a point that someone thought the boss should know about. The result was that the boss *did not* know that important point at all. At the case study organization, the assistant teacher interviewed noted – and this seemed to be a common perception when discussing with teachers – that students have become scrupulous in contacting teachers through e-mail. It is as though it has become so easy to ask the teachers questions that students bother them with the most trivial queries, queries that can be easily looked up on the web, a printed study guide or somewhere else. Also, because of mobile phones, teachers complained that students expect them to be "on duty" all the time and do not ever seem to accept no for an answer when they are told that they are calling at a bad moment. Yet the interviewees thought that the good side of having e-mail overcame such small negative side-effects.

There were several references to the medium through which discussion on workplace matters should take place. If the matter was sensitive, or regarded as a negative issue, it seemed to be easy to use e-mail to avoid direct confrontation. However, everyone did not agree that this was the best forum for such discussions but would have preferred face-to-face interaction, or preferred not to be involved in issues, as some discussions did not concern all recipients of discussion e-mails at all. These incidents could be interpreted as ICT changing the culture of how to interact with peers and superiors. The case of employees sending copies of mail conversations to superiors looks like a new form of tale-telling, supported by the ease of this medium.

Trustworthiness of information

Another interesting issue about how to handle information came up in some of the interviews: the difficulty of judging whether the information you find is trustworthy or not. This issue is, of course, of particular interest to the university and research world, but also to the general public, as the Internet is increasingly becoming an information-seeking tool for everybody with access. One interviewee (a woman in her forties) spoke on the subject as follows:

> Now anyone with computer and server access can publish information, so you need a totally new type of skill, because you yourself need to do the quality control of the information … can [with emphasis] you trust the things you find? … there are a lot of strange things as well.

Interviewer: Do you think it is difficult to learn this kind of skill?
Interviewee: Yes! [With emphasis.] I mean, for a person who has grown up in the way that at least what you read in the morning in *Husis* or *Hesari* [short for the two major national newspapers], well, at least that is – if not completely true – at least almost [with emphasis] the truth! So the fact that someone has produced a text, then you usually believe … or it is like our [business] school … that at least, I have been brought up to believe that what is said in the school book is true! [With emphasis.] And what is written in *Husis* is true [with emphasis] … and what Arvi Lindh says in the TV news is true [with emphasis] … because then there has been an authority or someone else who has controlled this truth. But now, then when it is about the information you find on the web, we need to go and look at the source … and … all this is very exciting!

She then goes on to describe how in the UK, for example, all academic e-mail and www addresses end with "ac.uk", and how this "ac" has become an implicit symbol for judging what kind of information can be expected on that page. She gives another example on how she accidentally entered the home page of the finance department of a competing business school, also state-owned, and found that they had advertisements there. In Finland and the Nordic region, most schools and universities are state-owned and commercial interests are kept outside. However, this is slowly changing with increasing sponsorship and advertising. She comments:

> Hanken hasn't started selling ad space yet, but somehow, for about ten seconds, I was thinking, "Damn, Helsinki School of Economics [the other business school] has started selling this … is it really sensible for a university to go in on this, that we sell ad space on our web pages, will it lower our trustworthiness or not?" I haven't thought about it since then, I just remembered it now … But really, how are you to judge the trustworthiness of the information on the web, and how do you build up this skill then?

The interviewee has already started developing her own new rules of thumb, and she is no doubt sharing and discussing these concerns with others, spreading these insights further.

Changing perceptions of time

Another source of confusion seemed to stem from the changing perceptions of time that ICT brought and of what this meant at the individual level for understanding one's work and how to perform it. Most interviewees and workshop participants told several stories on how e-mail had increased the pace of things: because it is so quick and easy it also increases the pressure to be quick yourself in answering and getting things done. The Internet seemed to have an effect on how time is perceived in relation to work and tasks.

An interviewee (a woman in her fifties) spoke about surfing on the Internet:

> As many people say that "now I have seen this and that and that" but, my God, when do you have time during work hours? I mean, an hour has passed quickly when you are in there surfing, so there really isn't time … and then one has this strange … or I guess I was brought up in such a way that, like earlier, some years ago, when the net wasn't so common, then you almost got a bad conscience from being on the Internet! If your boss passed by, you got this feeling that here I sit and just play around and don't do my job! [To the younger interviewer:] You have probably never [with emphasis] felt like that?!

Another interviewee (a woman in her forties) put it like this:

> There is not yet this … or we haven't really gone into this kind of e-mail culture yet. It's like this, that if you send an e-mail to someone it's almost like you have to answer it within two minutes – you expect this kind of fast answers. Earlier, when you had … or take for example ordinary mail, then it was okay and you could contemplate on your answer for a week and then send it back. I mean that we haven't yet really found a culture for how to answer or if one has to answer immediately or what.

The opposite idea was that e-mail gave people time to consider their answers to a question, as it can be seen as less obtrusive than calling on the phone. These kinds of comparisons can be interpreted as sense-making processes in trying to understand and learn how the media should be used and testing it out during interactions with others. Furthermore, the separation between work and leisure time is increasingly difficult because of ICT. This seems to result not only in increased stress, but in increased feelings of guilt for being too lazy, of pressure to work in spare time, and of general uncertainty and confusion over how to behave at the workplace. The pictures that computer and telecom companies try to paint – that ICT and mobile devices make your life easier and you perform your job faster – are only partially true. It seems that with technology the pressure and expectations on performance increase even more: with mobile phones you are able, and therefore expected, to answer your calls at any time of the day or week, or even during holidays, or with laptop, palmtop or communicator you are able, and therefore expected, to start working already in the car or bus or whatever transportation you take to work. The boundaries between work and leisure are becoming blurred, and being able to switch off work and be unavailable, dedicating time to home, family, friends or hobbies, is perhaps becoming the luxury of the future.

Discussions and conclusions

Comparing this analysis to previous research on ICT use and the relationship between organizations and technology confirms several things:

* once the ICT was in place, its use was flexible according to users' interpretations of it and changing interpretations as learning occurred, i.e. the technology was not static but remained interpretively flexible;
* decisions on technology introduction and use were based on single individuals' personal hunches, i.e. decision-making was neither always formal nor rational but could be quite coincidental, based on influences from anywhere in the surrounding context;
* knowledge was transferred from individual to individual like ripples on a pond, quite untouched by hierarchical levels or formal organizational structures, i.e. personal interaction was key for learning and gave a central role to those who had knowledge about ICT use they could share.

The term "virtualization" here refers to the iterative, micro-level processes between ICT use and learning and the institutionalization of this technology into the context of the user and the use, changing the initial use and understanding of the ICT. This chapter has addressed the new tacit knowledge and understanding that is collectively and individually emerging in organizations as virtualization proceeds. What is going on within the micro-level processes of virtualization could be explained in terms of a social learning and construction process. The process iterated between users, peers and enthusiasts, and the fuel at this level seemed to be the sentiment of getting individual satisfaction or dealing with a moment of frustration. Regulations and decisions made outside organizational boundaries also comprised the frames for the decision space within which organizational and individual choices of ICT and its use had to be made.

In summarizing the web of forces and processes sustaining and driving virtualization, influencing ICT use at both micro- and macro-levels in the organization, at least the following should be included:

* *technology enthusiasts* of different kinds, not necessarily experts but those who can decrease uncertainty for a user in a given situation;
* *emotions driving learning*, especially the frustration of being unable to solve problems in relation to using ICT and the joy and fascination of being able to do so;

- *habitual use of ICT* together with *features in ICT* that make humans dependent on it, yet leaving them with less control over their tools;
- *technology supportive culture*, including an admiration for technology and people who are knowledgeable about technology, with an often firm belief that technology will make things easier and better for humans, without having to prove itself;
- *political and other external forces* regarding structural, institutional and economic matters, such as infrastructure, national and international standards, laws and regulations regarding technology and its use;
- *commercial competition* between organizations.

One area where to some extent unexpectedly large effects could be detected was in the structuring of information. Uncertainty on information matters gave rise to continuous interpreting, re-interpreting and negotiation of how the existing ICT was supposed to be used and why. Other micro-level effects of ICT use included changes in communication patterns between employees and superiors or teachers and students. How to judge the validity of information and its source was an issue of concern for interviewees that needed both learning and further interpretation and negotiation of useful rules of thumb. The fast pace of e-mail interaction affected – not always positively – perceptions of time and the ability to concentrate for longer periods. The Internet seemed to affect how time in relation to work and tasks was perceived. The instability of new technical and socio-technical systems seemed to be a large source of insecurity, even duplication of work.

Virtualization is no longer really about ICT use, but completely new ways of acting and knowing things in an increasingly ICT-supported environment, both in organizations and elsewhere in society. Technology becomes the medium for producing other action; it becomes institutionalized as part of its context. It is about the ever-changing meanings and new interpretations of what ICT is about and for what and how it is supposed to be used. It is also about how technology reshapes social structures and behaviour, and about changes in power, roles, norms, values, rules and regulations, starting often in organizational settings but reflecting on to all other social spheres as well. It is about a transformation of the contexts in which work is done and life lived, and on how humans behave and communicate with each other in that transformed context. It is about how people adapt to new technology: how they make it their friend, struggle to understand its purpose and find joy in learning new ways to do things. This tells us something about the incredible power of the human mind when brought together as a collective force, and how humans slowly but surely are working towards a virtualized future.

Simultaneously, the speed of social construction and learning seems today to be such that individuals and organizations seem to feel they are lagging behind technological development. In a global world there is always someone else further along the line of development. This chapter has tried to show that micro-level drivers of ICT use and the consequences of this use are interrelated in that they are to a large extent driving each other. The relationship is not simple but iterative, and works across levels and beyond organizational boundaries. A need to compete and the belief in progress, both particularly characteristic of Western capitalist societies, act as fuel speeding the process. No longer do people and organizations compete with their "neighbours", but the competition is global, even at individual and micro-levels. Because the quest for being first operates on a global scale there is not much room for single individuals, organizations or even nations to critically question the reasons for and the ways in which ICT is implemented and used. In Western society there seems rarely to be time to evaluate real needs or think about further consequences of ICT use. New technologies are implemented immediately, while the consequences are problems of tomorrow, to be solved by someone else. Individuals and organizations alike fall victim to the constant technology race, without always understanding its future consequences.

Acknowledgements

The author wishes to thank the Finnish Foundation for Economic Education for their generous support in regard to this research. Thanks for helpful comments are due to Guy Ahonen and especially to Jeff Hearn and Anne Kovalainen for carefully commenting on various versions of the chapter.

Note

1   The word *beroende*, used by both the interviewer and the interviewee, means, in Swedish, both "dependent" and "addicted". In reply the interviewee played on the meaning of the word, changing it from "dependence" to "addiction".

7 Women and technological pleasure at work?

Päivi Korvajärvi

Introduction

Having done research since the early 1980s on women and technology at work, there is a recurring issue in Finnish white-collar and clerical workplaces that strikes me as important. It is the eagerness and even joy of women in using various information and communication technology (ICT) in offices located in various branches of industry. I have hardly met any women who have resisted new equipment or software applications. On the contrary, the women even seem to love their work with new computers and new software applications. On the one hand, the clerical workers find ICT to be a necessary tool at work. When, in interviews in the 1980s and 1990s, I asked clerical workers in different kinds of clerical workplaces the meaning of ICT they stated plainly, "Technology is a big issue here, we cannot do anything here without utilizing it." On the other hand, ICT was not only a necessity, it was also a source of enjoyment at work.

When computers were widely introduced in clerical workplaces in the early 1980s, the introduction somehow made clerical work more visible there. Clerical work got attention and its significance became visible, although the processes inherent in the introduction of the new technology were top-down processes, and usually directed by automatic data processing (ADP) experts. Still, the clerical workers welcomed technology in their work. The reason seemed to be that the technology gave the clerical workers new ways of thinking and learning in their work. I have never met a clerical worker who has wanted to stop using ICT in his/her work. Technology has in the longer term meant a crucial change in clerical work that has traditionally been invisible, unchanged and undervalued in organizational entities, although in the beginning the change process was slow (Korvajärvi 1990). Today the clerical and white-collar women use the technology fluently – in fact, the majority of women in the workforce use ICT in their work in Finland (Lehto 1999); however, in cases of disruption the women usually call men, who come to make the necessary repairs or give advice. In this regard, the main aspects of the clerical work have remained the same as in the 1960s. Crozier's (Crozier 1964) description of the maintenance workers' strategy to keep their autonomy in technical matters, seems to once again be valid today.

The prevailing argument in feminist studies says that women are more or less excluded from the making of technology (Webster 1996: 66). Accordingly, it is men who are powerful actors in the realm of technology. Men are developers, designers and rescuers in case of emergencies. Consequently, men represent the experts and forerunners in the visions for the future development of an information society. Empirical studies in Finland show a close cultural association between technology, men's dominance and masculinities at work and also in the other communities of technology use (Vehviläinen 1999a; Korvajärvi and Lavikka 2000). This contradiction has led me to question whether I have missed the obvious in the workplaces, or whether women simply enjoy and feel pleasure in using ICT at work, despite the fact that they are outsiders in the making of the technology. It is this question that is the topic of this chapter.

More specifically, this chapter aims to present a more complex view of women and their relationships towards technology at work. The main argument presented here is that women, in certain conditions, enjoy the use of technology. Moreover, working with technology, in certain conditions, is a pleasure for women. It is precisely the technology that makes the work and the work organization attractive for women. Accordingly, in this chapter I focus on women's positive experiences of ICT in work. This is what I call here "technological pleasure". What are the material and symbolic aspects of technology that women enjoy and that attract women in work? How does ICT provide pleasure for women in work? The aim is to analyse the ways in which women's pleasure in using technology "works" in a work organization, on the one hand, and the consequences of the positive association between women and technology at work, on the other.

In this chapter the concrete meeting point of women and ICT that is examined is the call centre as a new form of workplace. In the growing body of research on work in call centres there is a trend towards emphasizing the negative aspects of the job content. Thus the dominant interpretations find call centre jobs to be characterized by Taylorized routine (Bagnara 2000); reproduction of inequality between women and men (Poynter and de Miranda 2000); and total managerial and technological control of everyday work (Taylor 1998). It has been argued that in the call centre, emotional labour, as managerially prescribed behavioural requirements of the employees towards customers and clients, is a conscious managerial strategy for utilizing the female labour force (Taylor and Tyler 2000). According to some studies, call centres represent a pessimistic vision of work in an information society in which employees should and do work and feel like manipulated robots. However, there are slightly different views that challenge this picture of call centres. It has been shown that the employees in the call centre feel emotional ambivalence that includes both enjoyment of customer service and frustration with their low position in the organization (Pratt and Doucet 2000). In addition, a more sophisticated and more optimistic picture is provided in an international comparison of work in call centres. This suggests that versatile work content can also be a realistic alternative in the call centres. However, the most probable situation is that different features of work contents prevail side by side in call centres. The crucial issues are organizational choices made by the management (Frenkel *et al.* 1999; Callaghan and Thompson 2001).

The technological pleasure for female employees in call centres has not been examined previously. In contrast, technology and work organization in call centres has usually been interpreted as representing something negative for the employees, as mentioned above. Call centres are workplaces whose existence is impossible without the technological applications that integrate telephone and computer technology. In this sense call centres can be seen as "technology-driven" workplaces, and their services are dependent on technological solutions. In addition, the technology is a very close and integral part of the employees' work there. The employees wear headphones, and the computer and especially its display and keyboard are absolutely crucial elements that make the use of databases possible at work (see Poynter and de Miranda 2000).

Thus the material equipment is intertwined with the bodies of the employees in call centres. In addition, and perhaps more importantly, the material signs of interaction, i.e. in the case of the call centre, the voice and speech of the customers and the employees themselves cross the bodies only through technology. Thus the combination of the human bodies and the material technology shape the space for interaction with the clients and customers. This type of close linkage of the technologies and the bodies is a relatively new but increasingly emerging phenomenon in the labour process of the service sector. Traditionally, organizations have used technology to control the bodies of the employees; however, the close and concrete link between the bodies and technology can also produce emotional states, such as comfort and enjoyment, but also stress and anger (Fineman 2000: 9). The technology is material; however, the concrete findings that are presented in this chapter indicate that the symbolic aspects of technology bind the workers together and the symbolic aspects of technology are a crucial part of the creation of the social community of work. The technology functions powerfully through its symbolic meanings at work.

In the remainder of this chapter I discuss the major dimensions of concrete work and its symbolic content that lean on the material aspects of technology in call centres. The conceptual configurations are drawn from various sources such as: conceptualizations of job satisfaction and desire; research on technology and pleasure; and research on symbolic aspects of technological change in organizations. In this complex field, analysis is centred on the women studied as actors and the symbolic aspects of technology embedded in the women's activities. I present a variety of approaches in the literature that conceptually illuminate pleasure at work, and continue with a discussion of the particular call centre that was the focus of this research.

Women, pleasure and technology

No chances for technological pleasure?

Technology, pleasure and women at work is a topic that is rarely discussed in the literature on ICT and gendering processes. The terms seem mutually contradictory. In the background there might prevail an understanding that waged work, no matter whether it is women's or men's work, as such presents an inherently oppressive situation for people. Thus, from this point of view, enjoyment, pleasure and work are incompatible with each other. Technology and pleasure are seen as intertwined, however, only in the case of men, usually those working in engineering (Hacker 1989), or hitherto also carrying out tasks that are connected with virtual organizations, multimedia or cyberspace. The conclusion is that women are the ones whose pleasure is excluded from the ICT-mediated spaces.

In an analysis of the relationship of technology and gender inequality in computer education, a slightly different conclusion was drawn. Henwood *et al.* (2000) show that while women's access to courses on computing and information technology may not be formally denied, women may underestimate their technical skills and competence. In this way, women themselves could be said to reinforce the prominence of men and certain masculinities in the field of technology, such that the possibility that women would enjoy using technology at work seems even more unrealistic.

Pleasure and job satisfaction

It is necessary to ask for the meaning of the term "pleasure" in terms of women, work and technology. How should "pleasure" be defined? How does it differ from the widely used concept of job satisfaction, for example? It is the concept of job satisfaction with which I have chosen to start.

The central aim of the concept of job satisfaction is to recognize the motivational factors that increase the productivity of employees. Although there are several directions in studies of job satisfaction, there has always been a link between job satisfaction and work performance. The crucial point is how management develops job designs that promote effective self-actualization of the employees at work. The motivation of employees has a strict goal: that of efficiency and productivity according to the lines set by management. The perspective of job satisfaction in research has been the perspective of the management. It has been focused on the individual, whom the managers motivate to work, and the redesign is focused on a single job. The development strategy for the jobs is based on big surveys that originally included only professional and managerial people, for example in the studies of Frederick Herzberg, the founding father of job satisfaction (Hollway 1991: 98–9, 102–7; Grint 1998: 125–6). The job satisfaction approach makes the assumption that it is possible to direct the motivation, the drive of the employees for satisfaction at work, according to organizational needs. Thus social necessities shape the quality of motivation of the employees. Both the motivation to work and the goal to be satisfied with the job are defined outside of the "motivated" employees. Motivation and job satisfaction are found to be aspects of a cognitive process (cf. Thompson and McHugh 1990: 266–79).

The approach of this line of job satisfaction research differs from my approach in this chapter. My aim is to examine women's positive relationships to technology through the expressions of the women themselves, not through the conscious job design models developed by the managers. Thus, I understand pleasure as an orientation that expresses women's enjoyment of technology, their attraction to and involvement with it. In this respect pleasure is close to the definition of job satisfaction that refers to "a pleasurable or positive emotional state resulting from the appraisal of one's job or job experiences" (Locke 1976: 1300, cited in Frenkel *et al.* 1999: 231). However, technology as such does not play a role in that definition. Moreover, in the job satisfaction approach gender is used as an exogenous variable in the same ways as occupation is used, or the level of technology is used at work. In addition, the quantitative setting used in job satisfaction studies excludes gender understood as practical and symbolic processes (Acker 1992; Gherardi 1995). The latter suits the Finnish context well, because it is the contents of space at work that are important for women.

Pleasure and technology

Sally Hacker (1989, 1990) has explicitly analysed the relationships between technology, pleasure and work. She argues that technology and pleasure have strong bonds with each other at work (1989: 5–8). By technology she understands both machines and complex systems of social relations, including the design, the development and the use of technology, and the passions of people. Sally Hacker has suggested that the dimension of pleasure is embedded in the technology. According to her, technology includes "sensual and erotic dimensions", and it is even "appealing to sexual desire" (1990: 8). In her understanding, the erotic dimension of the technology is related to power and to powerlessness. Central in her thinking is the argument that technology consists of male power and this is why technology is seductive for men. This was clear in her case study of engineering education at MIT, where rational criteria are stressed. Rationality was connected to the ways of organizing – an ideal type of which is the military organization – "control of sensuality, emotions, passion, one's very physical rhythms" (Hacker 1990: 209).

Thus Hacker stresses technology as a tool for control and suppression in organizations. According to her, the control aspect combined with the hierarchical organizing of work attracts male engineers. Technology, power and men belong together and their association constrains women's chances and spaces at work. Thus the core conditions for changing the situation are autonomy and opportunities for collective activity at work.

The lesson from Hacker is that one possible path of looking for pleasure in technology at work is to examine social processes around the technology. Since Hacker emphasizes the rational aspects of technology in the sense of strict, formal, army-type organization, another lesson from her contribution is to look for aspects of technology that are non-rational in order to find women's technological pleasures at work. The former path leads to an analysis of the different types of interaction and communication mediated and performed through and with the help of technology. Thus the focus is on the concrete work content and especially on the autonomy to interact and communicate. The latter path of non-rational aspects of technology leads to an analysis of issues that are connected to emotions and commitment at work. Here the focus is on the ideas, thoughts and passions of the employees at work. The former path includes more or less visible aspects in a work organization, whereas the latter path includes aspects that are more or less invisible and even hidden. It is the latter path that gives more insight into technological pleasures at work. However, these two paths of analysis are not in contradiction or separate. Instead, they are intertwined with each other.

It might be the strong point in Hacker's thinking that she does not define technological pleasure as a strict concept. Rather, technological pleasure includes a variety of impressions, opportunities and emotions. There is one condition for defining something as pleasure for Hacker. Expressing and feeling pleasure needs autonomy in the sense of both physical and mental space. Hacker stresses erotic dimensions of technology. However, she also includes in her notion of pleasure opportunities to express oneself most fully (Hacker 1990: 5). She seems to use in a parallel way the terms "erotic", "strong feeling", "comfort" and "pleasure" (Hacker 1990: 220). What I'm interested in is to analyse the aspects of technology that give comfort and enjoyment to the employees in call centres. This is an elusive task. However, I include in these aspects not only the material organization consisting of interactional activities, but also symbols that activate emotions and commitment.

Pleasures, interpellations and subjectivity

John Law (1998, 2001) also uncovers the relationships between technology and pleasures. He speaks about "machinic pleasures" and uses as his theoretical stance the concept of interpellation derived from Louis Althusser. It is a question of the interpellative effects of technology that attract people. A good example of an interpellative impact of technology in the Finnish context is that technology is found to be a sign of well-being and a competitive edge in the globalizing market. In fact, in the current Finnish culture, technology is a prominent or even sacred feature that seduces everyone, or at least does not leave anybody unchanged (cf. Althusser 1971: 160–70). It is not necessarily technology but a well-being guarantee through the use of technology that attracts people. Althusser was interested in how ideology constitutes concrete individuals as subjects (1971: 160). As a research strategy, looking for subjectivities through interpellations means reading that includes absences and silences in the texts or other research material. Following Althusser, Law deals with the question: how does the technology, through its interpellative effects, create certain kinds of subjects in the specific contexts?

Law (1998, 2001) suggests a selection of machinic pleasures. All the pleasures are specifically male ones. Thus, the approach is consciously gendered and the strategic question is what kind of pleasures the technology – in his case the technology used in the aircraft industry – creates as male pleasures. The question concerns how male subjectivities are created through military technology.

His narratives relate machinic heroism, which includes pleasures of male camaraderie and homosociality in which fighting and death are part of the bodily games, so that the sense of time disappears. Virtual combat refers to the games in which both "the reality and the unreality of combat" (Law 1998: 28) are reflected. The games may be understood as technology that gives pleasure, or at least limits pain. Passivity describes the pleasure of routine activities in which there is no space for an acting subject but only for the subject that absolutely relies on the technology – during take-off, for example. Further enumeration tells about the absence of relationships between things, events or objects. Pleasure is "in recognition of something that is, that is out there" (*ibid.*: 38). Moreover, completion includes the idea that technology is autonomous and that as such it functions perfectly. Consequently, the pleasures come from its ability to control and to maintain itself. The pleasure from the extension means that one's body is extended with a tool or machine and the boundaries between them are obscure. This is close to the description of the role of the technology in relation to the bodies of the employees in the call centre work. Finally, beauty is an aesthetic relation to the material technological object (*ibid.*: 24–45).

Such pleasures have erotic content and they are related to autonomy and control in relation to the activities of people. The situation is complex, however: the pleasures create male subjectivities that bind men and technology together locally in specific ways. In my reading, the pleasures suggested by Law can be comprehended also as symbolic aspects of technology functioning in the processes of creating subjectivities. In the same way, there is a question what kind of female subjectivities technological interpellations are able to create.

Technology and symbols

Symbolic processes underlying technology and technological change have been under consideration, although relatively rarely. Puskala Prasad (1993) found a variety of symbols that technology represented for the employees she studied. She does not inform the reader whether the employees in her case study organization of a health maintenance centre were women or men. However, I assume that women were well represented among the employees, because nurses and receptionists made up nearly a half of her interviewees. In spite of the gender-neutrality in her writing, Prasad's study gives fruitful insights into the symbolic dimensions of technology at work. She understands technology as the process of introducing new computer systems in the case study organization, and in this process she focused on the symbols of computerization both before and after the process. Prasad found three categories of symbolism connected to the computerization: the pragmatic, the pessimistic and the romantic.

The pragmatic representations were related to organizational survival. The computers were perceived as improving efficiency and professionalism, and computers were seen as inevitable. The pessimistic representations included negative aspects of computers, like their omnipotence, that would lead to hazards, organizational turmoil, robotization and otherness. The romantic representations were more or less idealistic. They included the symbols that give human characteristics to the computers (anthropomorphism). Furthermore, the romantic symbols consisted of far-reaching expectations (utopianism), and of excitement at the chances to have fun and to play. The pleasure and enjoyment aspects were mostly represented in the romantic category. In addition, the pragmatic aspects included positive orientations although they were mainly instrumental in relation to the work processes.

What is provocative in Prasad's study is that on the local level, for different members of the organizations, the meanings of the categories varied. In addition, the representations varied according to the phase of computerization in the organization. For example, professionalism as practical symbolism meant that computerization made the organization cleaner and more organized instead of having a rather messy appearance. In addition, computerization meant that technology would raise the organization to a more developed level and transform the employees into more modern experts. Furthermore, human aspects of computerization also had several different contents. It represented an indivisible association between technology and the human mind and between technology and intelligence. Computerization was seen as superior brainpower. Moreover, computerization could also be connected with humanness, emotionality, affection and warmth. The human characteristics of the computers, in terms of both brains and affection, were pervasive among all members of the case study organization. Both expectations of the development and human aspects of the computers became sedimented in the organization so that the employees from different groups used them as their resources in their daily work (Prasad 1993).

The processes had major consequences in the organization. The symbolism of professionalism, including its positive expectations of future work, created a positive climate for computerization, excluding resistance towards it. In addition, the symbolism of professionalism provided status and thus ensured a long-term commitment by the employees to computerization. The symbolism related to the human aspects made the computerization glamorous and trustworthy. Because computers had an image of human intelligence, in cases of human error the computers were blamed as stupid (Prasad 1993). It is possible to put Prasad's line of thought into the context of Althusser's interpellations and Law's examples of it. Human characteristics of computers, far-reaching expectations or excitement are interpellations that the technology has made possible or offered to the people in certain conditions. Consequently, the interpellative processes offered space for subjectivities that included both glorification and scepticism towards technology at work.

Comparison among approaches

Hacker's, Law's and Prasad's studies have similarities, especially in their methods and in the targets of the research. All of them use ethnographic methods in data collecting. Hacker calls her method "being with people", whereas Prasad leans strictly on symbolic interactionism looking for meaningful activities and processes, and Law constructs narratives. Perhaps Prasad and Law have done their fieldwork and analysis in a more organized and conventional way compared with Hacker; however, all three authors are looking for local practices in relation to technology use or to its implementation. The target of the authors is to account for the diversity and variety of everyday life at work. Hacker and Prasad deal with the concrete and symbolic meanings of technology. Prasad stresses the influence of symbolic representations on the concrete flow of everyday work and its changes, whereas Hacker's line of thought goes from material to symbolic aspects. Law in his analysis integrates technology and the activities of the people in the aircraft industry and looks for ways in which technology and subjectivities are locally intertwined. All three authors see the material and symbolic dimensions as inseparable in concrete work.

The job satisfaction approach partly deals with the same issues as the stress on autonomy and self-actualization that are found to be preconditions for motivation and productivity. However, the job satisfaction approach lacks the erotic and emotional dimensions, as well as the technological. The most prominent feature in the job satisfaction approach, compared to Hacker, Law or Prasad, is the understanding of job satisfaction, and in this sense (technological) pleasure at work, as comprised of variables that are universally applicable. In addition, the job satisfaction approach presupposes certain concrete motivational factors that lead to the best possible productivity rates as determined by the management. The other theoretical conceptualizations referred to here take processual or situational stances towards empirical reality. Pleasures are different in different contexts and they are processes in which bodies, emotions and erotics meet each other and, in this case, interact with technology. Thus the approaches suggested by Hacker, Law and Prasad have connections to the conceptualization of emotions at work (Fineman 2000); embodied workers in organizations and bodily performances in work (Hassard *et al.* 2000); or sexuality as it is intertwined in the secretarial work and the construction of technology use in that work (Pringle 1989). However, I find the perspectives from Hacker, Law and Prasad interesting enough. Thus the discussions focused particularly on emotions, embodiment and sexuality are beyond the scope of this chapter, and the emphasis is on the contributions that keep technology on the foreground in the examination.

Technological pleasure can both empower and disempower people, whereas the goal of the job satisfaction approach is to produce people who work effectively. Thus the theoretical landscape of the job satisfaction approach is different from the other contributions examined here. The difference appears particularly in regard to how the context of the study is constructed in terms of the relationships between technology and those researched on the one hand, and the relationships between the researcher and the researched on the other.

Hacker's and Prasad's inspiring studies had been completed before the availability or even the existence of the kind of communication and information technology that is currently used. The studies were conducted at a time when even computerization was a new issue, and not as self-evident as today. Thus, in contrast to Hacker's and Prasad's cases, it is to a much more advanced workplace in terms of technology that I now turn. In contrast to Law's case, women in a female-dominated workplace are the ones whose pleasure I am interested in.

The call centre as a place of work and enjoyment

The case study call centre is an unusual place because applications are not only bought but also developed there. In the case study call centre the new software and other ICT solutions and new services that the company offered to its customers, in the context of both business-to-business and customer-to-customer services, were closely intertwined with each other. Usually there was first an idea of a new service, and then the appropriate ICT solution was designed. After that the concrete work arrangements for the new service were launched. Thus the design process of the ICT solutions already included the design of social practices in the performing of the new tasks. The organizational forms took on concrete form in the development process of the new service. Thus the material technology as such did not direct the social organization. When thinking about ICT as material machines and channels, the design process was crucial in building the concrete relationship between the ICT and its user.

Because I preferred to examine experiences, processes and situations of technological pleasure at work, I chose a loose ethnographic approach in the fieldwork. Thus the research material is qualitative and it mostly consists of semi-structured interviews. The aim was to cover in each interview the following topics: work history; changes at work; the demands of the job, especially in regard to customer service and ICT use; the employees' degree of responsibility in the job; control at work; co-operation and autonomy at work; competition; occupational identity; participation; gender aspects of the work; and visions for the future. The interviews concerned the facts as changes at work but also the employees' experiences of the changes. In addition, I had opportunities to observe the activities of the employees in the call centre and to use the headphones and listen to the interaction between the employees and the customers. I also gave some feedback and in this way tried to create an interactive research relationship with the employees.

I conducted the major part of the observations and interviews in the case study organization (twenty-nine employees) in September and October 1998. The fieldwork took eight days and I conducted observations and interviews during those days. Furthermore, I reported the findings to the employees in November 1999, and we discussed my interpretations. In addition, I conducted three interviews in November 1999. I visited the call centre and carried out five interviews in June 2001. In the analysis I have sifted out of the research material the experiences, accounts, feelings and opinions that I have interpreted as implying technological pleasure at work. Thus the aim is to make visible a phenomenon that is usually passed over in the research. The analysis is based on looking for facts and subtexts that I find to represent technological pleasure. Consequently, the written ethnography is not only the result of the fieldwork. It is also, and even more so, the writer's interpretation of how technological pleasure is concretely and culturally present in the case study organization (see Van Maanen 1988: 4).

In the call centre case study organization, the designers were men, whereas the huge majority of the customer service representatives were women. In addition, when the service was considered as "technical", for example the new service of the IT helpdesk, the call centre company hired mostly men for those jobs. The services that included using ICT as a necessary part of work and communication with the customers but did not concern any IT-related matters were women's tasks. As a female supervisor said: "In general, the girls in the call centre are not very much interested in technology. They just use it. Those who work in the IT HelpDesk must be interested in the device, how it works." The boundary between women and technology was in speech. If the customer service representative (CSR) had to speak about technical matters with the customers, then the jobs were coded as if they were men's jobs. Thus the case study organization followed the model, also as a social situation, in which men are associated with ICT and women with services.

The issue of how CSRs in the call centres themselves relate to the technology is a research topic that is nearly untouched. Researchers have studied the control dimensions embedded in the software; however, the subjective experiences of the employees have not been reported. Some evidence on the pragmatic type of orientations has been reported. In a comparative study it appeared that the effectiveness of technology is more important than the opportunity to control technology for the CSRs. Especially, the CSRs in service jobs compared with sales or expert jobs (like IT experts) appreciated their overall effectiveness rather than just the idea of having control over technology. The relationship to the technology was instrumental. The other side of the coin was that, in general, this control and its means were beyond the control of the CSRs (Frenkel *et al.* 1999: 87–8, 242).

The high turnover rates in call centres have been used as an indicator of oppressive situations. The turnover in the case study call centre was nearly zero. The CSRs had full-time, non-fixed job contracts. The employees' ages ranged from 25 to 60. Compared to many studies, the employees had come to the call centre with the aim of working longer there. The call centre did not offer them just a workplace among many options, or a temporary workplace for students or housewives. On the contrary, even young employees said that it was quite possible that they would be doing the same jobs in the same call centre after five years, or even until retirement.

Thus the CSRs seemed to be deeply committed to their jobs and firm. On the one hand, this was due to the fact there were not a lot of job options for women who had usually completed a two-year commercial college degree after high school. The call centre was a workplace in which to earn a regular but low salary. The employees valued pay in the same way as did the CSRs in Australia, US and Japan (Frenkel *et al.* 1999: 61). On the other hand, the development of the case study call centre showed that it was safe to commit to the firm: it had started with four employees and in a couple of years the number of employees was over forty. The call centre had a short but promising past, and the future looked bright, with the plans for new services. Thus the progress of the case study call centre reflected the studies that have been done in Europe, although statistics on call centres in Europe are far from reliable (Bagnara 2000). In any case, it has been estimated that the number of call centres will increase dramatically in Europe. For example, in the UK in 1998, the number of the call centre employees was larger than the number of employees in the mining, iron and steel and car industries put together (http://www.labournet.org.uk/1998/August/call.html). This development is easily seen as a result of the development of a certain technology, although the development is linked with the growth of the service sector and with the trend of outsourcing service work.

The tasks in the case study call centre appeared to me as a researcher somewhat routine and relatively simple. At least, complexity and versatility of the work content were not my first impressions. Also, the CEO and a supervisor were worried that the work content was too boring and in this sense stressful. The CEO stressed that "our product is service on the phone. This is why the human being is important. E-mail and Internet are only tools ... The human being who is answering the phone is the most important." In the interviews, the CSRs followed pretty much the same lines. They felt that the pattern of work was very similar from day to day. "You are always taking calls," as an employee told me. However, after saying this she followed – as also did the other interviewed employees – by saying that "every phone call is different. This is not boring and repetitive work ... This is rather versatile work." When I later asked in an interview whether she needed creativity in her work, the same person said: "Most of the time you are like a machine, so you only do this work." The interview is contradictory: the experience of the job content is both repetitive and versatile. According to it the job content includes both discomfort and pleasure in doing the job. The discomfort is in the talk linked with the machine-like routine, and in this sense with the technology, and the pleasure is in the social content of the phone calls.

The pleasure in the social content of the job consisted of several elements. First, different customers as such made the everyday work multifaceted. Even though the task of the CSRs was to ask the same questions, one after the other, the customers' replies varied from person to person. Even though the supplier had given model answers to certain questions, the employees varied their words.

The CSRs were also sensitive to the tone of the voices of the customers and they varied their own tones of speaking. On the one hand, "one can hear immediately from the voice whether the customer is cheerful". The way of speaking was changed also according to the situation. "I carefully listen to the tone of the voice of the customer. If it is very formal, then I also speak formally, and say what we have been told to say." On the other hand, "we speak to the customer in a rather different way from the way we speak at home," as a CSR said. Thus the employees evaluated the state of mind of the customer and behaved according to their evaluation.

The pleasure taken in direct customer relations described above came from the contacts with different customers. The fact that CSRs also took pleasure in their own performance meant that a variation in the customers' tone of voice had impacts on the CSRs' ways of responding to them. The basic line was to follow the customer and to adapt to the behaviour of the customer. It was performance that was unautonomous because CSRs let the customers direct the contact. The customers gave them the content of the activities; however, the customers were not aware of it. Pleasure consisted of self-controlled activities that at the same time aimed to control the customer. This is to say that the CSRs directed the customer to behave in a proper way in relation to the firm by getting him or her to order something during the phone call, or by creating an attractive image of the firm for the customer, for example. This is exactly the phenomenon that has been called "emotional labour". The consequences of emotional labour have usually been found to be negative, causing stress and a danger that the employees lose the authenticity of their own activities (Hochschild 1983: 187–8). However, in the case study organization, the CSRs enjoyed performing and adapting their activities to the customers' behaviour, without customers demanding them to do it. The CSRs coped with the assumed wishes of the customers and with the wish of the management to perform as "human beings". At the same time, they felt some discomfort with the routines of the job; however, they did not feel stressed because of the wishes of the customers or the management. Rather, the CSRs enjoyed fulfilling their explicit and assumed wishes.

Second, the various co-operation forms that were needed at work when communicating with customers and colleagues made the work content non-routine and pleasant for the employees. Although the CSRs were in personal contact with the customer and the job in this sense was done alone, every interviewed CSR said that they do their job together with other employees. Indeed, when problems came up during a phone call, the CSR could make the phone mute and ask advice from a colleague. This also needed particular skills. As a CSR reported: "One must venture to ask advice. Nobody can think, after doing the job for two weeks, that one knows everything." Very often the problems seemed to become clear after a talk with a colleague. Thus there prevailed a social community of work in the call centre, and in cases of problems the CSRs were dependent on the social community. It was significant in doing the job successfully. In addition, the CSRs found that tolerance, the skill of keeping calm and the finding of "the proper style" were important requirements for the job. "The proper style" included keeping a distance from the customer.

The CSRs used the social community of work as a resource in order to do the job in a proper way. It was not a resource in the sense that it could only be used when in contact with the customers. The social community of work included small activities, chatting, joking and discussing, which made the social atmosphere enjoyable at work. The importance of the social community of work and its significance is an issue that did not appear clearly in the CSR jobs in the call centres in Australia, US or Japan. At the same time CRSs in those countries very much enjoyed customer relationships (Frenkel *et al.* 1999: 70, 88–9). The pleasure of the social community came from the chance to discuss equally with colleagues both problems at work and other things in life. There was no need to act in a special way with colleagues, as was the case with the customers.

Third, the employees highly appreciated their managers and supervisors. They found the CEO to be a democratic leader who, as a CSR said, "wants to be close to us and likes to co-operate with us". Further, the employees felt that the managers and supervisors were very easy to approach. The employees felt that they had the freedom to make suggestions if they had them. The easy-going relationships among the employees in different positions was a pleasure for the CSRs. It also made the hierarchy invisible and gave a feeling of a social community in which equal people do their work. However, the equal, face-to-face situation with colleagues did not mean equal opportunities for women in terms of access for the female CSRs to the decision-making processes concerning either technology or work organization.

Fourth, the employees felt that the vision of continuous growth offered them chances to learn new things and to develop themselves. The female CSRs appreciated the workplace since "you never know in this firm what will happen next week" or "something is going on here all the time". The new ICT was a part of it; however, it was a kind of subtext when the employees were thinking about learning new things. A male CSR explained to me that "this is a branch of the future … After five years this firm will be three times bigger than now." A female supervisor said: "Everything is very exciting. We all have good opportunities. The number of call centres will grow in Finland and they will have a bright future. The human resources are the most important issue."

There was a mixture of optimistic expectations. Optimism was always expressed as a source of enjoyment. This is why I here interpret optimism as technological pleasure at work, even though it could be both realistic and romantic. An inevitable expectation was the constant growth and the consequent interesting and changing work content. It seemed to be self-evident for everyone, without a need to reflect more accurately the existing or potential markets. In addition, the employees had opinions that were based on a certain type of concrete activities. They raised excitement and created an unanticipated break in the usual routine. They were something that brought up new ideas and a feeling of being involved in new issues. This pleasure came directly from the use of the particular technology that brought with it the vision of growth, and consequently security, in the labour market. In addition, the technology and its changing usage in the services sector made the job even more exciting.

Three of the aspects of technological pleasure at work were related to the social relationships in different directions within and outside the firm. One aspect of technological pleasure among call centre employees was related to the chances of developing and changing themselves as individuals. ICT and its characteristics were in the background, on the one hand, and in the middle of the stage, on the other hand. It was in the background in the sense that the employees seldom mentioned ICT or technology. They thought that the use of technology was easy. "Technology comes as self-evident, this is customer service," as a CSR said. The only problem was the situations where the computer systems stopped working, or when there were disruptions in them. They thought that ICT was necessary as it was; it was self-evident. Technology as such was not questioned but, on the contrary, was admired. In this sense, technology took centre-stage. The relationships to other people and to the development of oneself were framed by the technology. The pleasures were mediated by the technology.

Thus, the positive relationship to the technology was very concrete and practical. At the same time, technology was something that not only guaranteed the existence of the firm and the jobs, but was also like an abstract invisible web that illuminated the firm with its promises. In this sense, both the concrete and everyday activities and the distant but positive future were mediated by the technology. Furthermore, technology acted as an intermediary, both very concretely through the headphones and computer and very abstractly and independently because of being self-evident. Its existence as such guaranteed social support, equal relationships between people in different positions, a safe future and chances for individual development at work. It bound together the social ties from different directions. The concrete relations mediated by the technology needed the active involvement of people, whereas the abstract and self-evident vision consisted of passive expectations.

It has been suggested that the commitment of the employees includes involvement that takes on moral overtones and that "extends the satisfaction of a merely personal interest in employment, income, and intrinsically rewarding work" (Lincoln and Kalleberg 1990: 22). Clearly the CSRs are committed to the organization. One reason is very pragmatic: they did not have many other choices in the local area. However, their commitment also included other aspects. I would argue that the main reason for the commitment was their pleasure and enjoyment of the social relationships in the everyday job. Although there was a lot of routine, there were new things that attracted them to and even created seduction in the job. The technology gave the opportunity to feel safe and exciting at the same time. The commitment leaned on the abstract promises that the technology in the particular firm was gaining.

Concluding remarks

The contents of the pleasures were multifaceted, including both empowerment and disempowerment. The women related to the technology as something self-evident and necessary that had provided them a job. In addition, the technology had offered them a job that included a variety of social relations. Partly, the social relations gave an opportunity for equal relationships with colleagues and with superiors. Thus, the technology mediated social relations which empowered women in the call centre under study. The empowerment concerned horizontal relationships, since the improvement of skills and the confidential discussions of the emotional aspects of work were possible. Simultaneously, in vertical social relationships the equality concerned only face-to-face relationships; however, it did not open up equal opportunities for women at work.

In customer relationships, the technology mediated interaction. The situation forced the CSRs to make distinctions and interpretations based on voices and to draw conclusions based on what they heard on the phone without seeing the partner with whom they were discussing. However, there is perhaps a more important aspect in the social relationships with the customers. The CSRs represented their firm to the customers and they had to direct, and in this sense to control, the interaction with the customers. The CSRs had to keep the customer relationship a pleasant situation and they had to convince the customers to remain loyal to the firm. Thus the women, through the technology-mediated relationships, exercised control on behalf of the employer. This type of control is called adjunct control (Tancred-Sheriff 1989). In the case of the CSRs, the technology provided an extension of their bodies and the interaction forced them to function as adjuncts to the goals set by the management. However, in light of my interviews and observations, there is no evidence to support the argument that the sexuality of the CSRs had been used as a tool to control the customers. Rather, it is possible to say that the control over the customers gave the CSRs chances to lean not only on their own personal skills, but also on the institutional power of the firm.

The chances to learn new things were direct impacts of the expanding ways of using technology and solutions for providing new services. The perspectives offered by the technology extended the visions of the activities at work and in the labour market in general. The new services that became available through the technological solutions and the expectations of the new services committed women to the technology. The CSRs considered the technology as the solution and the agent that would solve their future at work. Thus, the technology represented the decisive agent in their reflections of their future income and well-being.

It appeared that women's technological pleasures in the workplace are not directly connected to women's place in the work organization. Women enjoy their jobs even though the job is low-paid and on the lower ranking levels of the organizational hierarchies. A good position or the prospect of a rising career in the organizational hierarchy is not the issue when thinking about technological pleasure at work. The experiences of pleasure come from other sources. The technology as material equipment is one issue. However, it is not an issue as such. The crucial point is what kind of visions for the future the use and the continuous learning of the ICT provide for an employee. The visions could be very concrete, such as a permanent job contract. In this regard, the technology appeared to guarantee future prospects for the employees. The visions can also be symbolic and based on abstract faith, such as the challenge of taking part in the new development and being proud of it. The social relations, not only with the customers and the colleagues but also with superiors, are another source of pleasure. The technology symbolically framed the possibilities of pleasant relationships at work.

Hacker's (1990) suggestion of studying social processes in order to find technological pleasure appeared to be fruitful. Moreover, conceiving of technology as a mediator in the social space that the CSRs used at work proved to be worthwhile in the research. Unlike Hacker, I found it as difficult to interpret the social processes as having erotic implications in terms of the women's exercise of power. ICT was seductive for the women in the sense that it represented an independent future for them by providing jobs and consequently their own money. In this sense, ICT also embedded an element of power that at the same time guaranteed the CSRs independence from the welfare state, spouses or families, and the chance to earn a living for themselves or their families.

Thus technology had interpellative effects when mediating social relationships and representing chances for development. The effects both produced and reproduced pleasures connected with technology, as suggested by Law. However, compared with Law's findings, women did not develop machinic heroism; rather, they aimed to create equal relationships in the technology-framed situation. Playing games on the computers was very rare. Instead, passivity and enumeration in the sense of relying on the technology was present among women, as was also the extension of their bodies through technology. The work environment, with its elegant design and colours, surely presented aesthetic enjoyment. Compared with Law's suggestions, the interpellative effects of the technology in the call centre were more symbolic and distant than in the aircraft industry. The technology was not as serious for women as it seemed to be for men in certain conditions. Rather, the technology shaped mental states that made both the concrete work and the future look positive and bright.

The symbolic representations of technology suggested by Prasad (1993) imply a possible element in the interpretation of technological pleasure of women in the centre work. In her terms of symbolic representations of technology, the CSRs demonstrated professionalism as the potential of learning something new through ICT. The CSRs showed also pragmatism when they thought that ICT would guarantee the existence of the firm. Furthermore, the technology represented for them romantic utopianism in their belief that ICT would guarantee a secure and pleasant future. Interestingly, the CSRs also recognized aspects of the pessimistic representations suggested by Prasad (1993); however, they seemed to be insignificant in comparison with the comfortable representations of ICT.

When focusing on technological pleasure, it is fair to ask whether it is politically correct to speak about women taking pleasure in technology, since the use of the advanced technology generally takes place in hierarchically gendered organizations. It is not my aim to neglect the powerful conceptualizations of emotional labour and its negative consequences in call centre work (e.g. Taylor and Tyler 2000). However, my argument is that ICT also represents comfort, pleasure and enjoyment for workers in call centres. Moreover, the argument is that technology also represents pleasure for women, even though they are not considered as experts in technology. Thus, technological pleasure also has connections to the everyday social activities on the shop floor without a repressive element of power.

The concrete visions of ICT are local ones. They are connected to a certain organization and its economic or similar success. Still, the symbolic visions are a part of the general discussion and cultural orientation. They are similar to the dominant trust in the power and good of ICT use. In this sense, symbolic visions merge with the programmes carried out by governments and companies throughout market societies. From this point of view, women in a call centre can commit themselves to the same ideals along with transnational companies and associations of nation-states. In contemporary Finnish culture, technology is a prominent, even sacred, feature that seduces everyone, or at least does not leave anybody unchanged. Technological pleasure remains intertwined with women's activities and orientations at work.

8   Fulfilment or slavery?

The changing sense of self at work

Riitta Lavikka

In this chapter I examine what kinds of new orientations and processes the informational intensification of work, intertwined with organizational and cultural changes, creates in the inner mental space of people working in traditional manufacturing industries. I focus particularly on two case study companies within such industries: clothing and engineering. Changes occurring in the world of work affect people's everyday activities and alter relations among people. The nature of work is changing from material processes and functional division of labour towards processes that are information-intensive, discursive and highly integrated. In the analysis I have found it useful to re-read Durkheim's (1964) theory of the development of the division of labour from the present perspective of flexible and discursively organized production. One of the implications of the post-industrial nature of production is the dissolution of traditional mechanical social solidarity dominant in Taylorized mass production, and the development of organic forms of solidarity that give more space for individual consciousness needed in co-operative and integrated work organizations.

Another point of departure in the analysis is the concept of *Dasein*, "Being in the world", which is at the centre of Martin Heidegger's (2000) phenomenology. The understanding of self that this metaphor describes starts from the situation of each human being as thrown in the world. The self is an existential project embedded in action in concrete environment, a project which is realized during the life course for better or worse, including the forms of *Dasein* that women and men develop at work.

My interest in self-processes has a history. As a student in the mid-1970s I worked for the local union organization of metal workers in Tampere, the "Manchester" of Finland. It was a time when the local unions had a lot of influence in the factories, and political life in the union was lively. The metal workers' office was a busy headquarters of union activities. As an office secretary, my job was to do all the paperwork regarding union fees and unemployment benefits, pay the bills and act as bookkeeper. I also participated in the local union meetings, taking care of the minutes. During the meetings I became very familiar with the world of union activists and came to recognize the power they had, not only among organized labour but in the community as a whole. Questions arose such as: "Shall we stop the work to show solidarity with the Chilean workers for the whole day or for two hours? How do we react to the large redundancies in another city? Do we accept the employer's offer or shall we continue the pressure?" It was the discourse of power and responsibility. I regularly met all the union representatives of the factories who took care of the union tasks in their work environment. Shortly afterwards, the structural decline of redbrick industries began, and changes came with increasing speed. Almost thirty years later I met some of these people again, working as a researcher on shop-floor techno-organizational change processes of some of the remaining factories.

Although the reunion was mostly happy, it was often confused too. We made a passing reference to our mutual history, with a short glance laden with understanding, and, after a few jokes, with an embarrassed ironic smile we turned to concentrate on the current situation. It was a different world now; we were different people.

As for clothing workers, I have a close relationship with them too. After graduating from a school of journalism, I worked in working-class papers, including the publication of the textile and clothing workers' union. From this reporter job I was invited back to the university to work in research projects (Lavikka 1992, 1997, 1998). The years in the union paper were at a time when teamwork in industrial production was started, at first as occasional pilot projects, then gradually in more pilot projects than I was able to follow up through my reporter's work. At this point I moved to researching the phenomenon more closely. As a reporter, I had previously visited the clothing company in question and met the seamstress whose interview is part of the material analysed here.

Individual change processes, such as the changing sense of "self", are a relatively new area of interest for sociologists of work, too. Without this dimension of self-processes, it is impossible to understand post-industrial working life, characterized as it is by knowledge intensification and rapid organizational changes. The old central concepts of the sociology of work, such as control and resistance, working time or space, do not seem to be able to capture the current experiences and meanings people give them in workplaces. Neither has postmodernism been able to make sense of the transformation of social practices at work which continue to produce social and personal consequences in the life-worlds of women and men (Casey 1995, 2002; Julkunen 2001). The meanings of concepts are changing, and the old categories and concepts move like zombies, dead but still alive, in our analyses (Beck and Beck-Gernsheim 2002: 203). For instance, an employer's control over workers is replaced more and more by workers' self-control supported by their feelings of commitment and responsibility, true or acted. Also, the boundaries between work and private life seem to get blurred. Examples of our data show, for instance, that people at all hierarchical levels in companies take work home from the workplace. For managers and upper white-collar workers, this has long been a common practice, but currently many shop-floor workers say that, in addition to their working hours, they spend some of their own free time in studying work-related information. Keeping oneself up to date with the newest developments at work is currently considered to be workers' personal responsibility and they are expected to invest their own time and effort in it. The work situation often is structured by the hectic performing of tasks with no time to go into learning new things more deeply. Therefore, the shop-floor workers feel that they must, for example, be prepared to take the manuals of new programmed machines or disks home to study in more detail the new programs they are supposed to use in their work. Or they are expected to update their automatic data processing (ADP) skills on their own initiative during their "free time".

In the "good old days" in the 1970s, metal workers were proud of their craftsmanship and manual skills. It was a source of self-confidence at work and also a guarantee of the employer's and co-workers' respect. The best craftsmen used to be elected as union representatives. Their skills comprised their resource for union politics, and for resistance, too, if needed. Today the notion of resistance is not to be met very often on the shop floor. However, while it is something that still is there as an idea to be taken out into the public in extreme conflict situations, it no longer seems to characterize the working atmosphere, employer–employee relationships or everyday working attitudes. The acts of resistance are more like rituals, lacking actual power or pressure (see Beck and Beck-Gernsheim 2002 on the zombie category), but which need to be performed according to the local union tradition in certain situations. In contrast to the traditional resistance culture, co-operation and willingness to learn are the slogans of the day, and various technical skills are a self-evident precondition for the work. "It is the new style now. And it would be impossible otherwise," I was told in the case study factories.

The understanding of the techno-organizational environment in current workplaces and the growing demands at work are crucial in order to understand the mental self-processes of people, as well. As a way into this, I first provide some background context. Following our company survey during 1998 and 1999, the manufacturing companies in the Tampere region have been characterized by both the advanced use of information and communication technology (ICT) and the early restructuring of organizational forms (Schienstock *et al.* 2001: 27–45). Our fieldwork results, based on seven manufacturing case studies in the Tampere region, confirm this. Practically all the companies were early users of ADP and had already experienced several waves of ICT development. In four of the seven manufacturing case companies, sophisticated integrated ICT systems had either started recently or were underway. Thus the often heard rhetoric of Finland as an advanced information society has an empirical basis.[1]

Historically, the work organization of the Finnish engineering industry represented the system of so-called flexible Fordism in which individual craftsmanship was able to maintain its valued position. Industry mainly produced more or less tailored machines and equipment for other industries. Thus, the Finnish engineering industry was not able to make use fully of the benefits of mass production, and therefore was able to preserve a path to flexibility in the late 1980s and 1990s. However, in other manufacturing industries, such as the clothing or textile industries, the concept of Taylorized mass production was the dominant production concept during the 1960s and 1970s until the mid-1980s, when flexible production of short, quick-delivery product batches replaced mass production, and team working in a few years took over Taylorized working, which was becoming inefficient and obsolete (Lavikka 1997).

Technological and organizational reforms started in Finnish manufacturing companies earlier than was the case in Central European countries. The reasons for this were the successive economic crises that compelled Finnish industry to start restructuring as early as the late 1970s and early 1980s. The crises were the oil crisis in the 1970s, the collapse of Soviet exports, and the deep economic recession of the early 1990s. At the organizational level, active internationalization took place in the companies. By buying production facilities abroad, the companies both secured their position in important markets and bought new advanced technological expertise in their branches all over the world. The product concept and the functioning principle of the companies were customized through the "just-in-time" concept. Units were downsized. Gradually all other processes but the crucial core know-how of the companies were outsourced to subcontractors and system suppliers. Profit centres were formed around core processes. Flexible and lean organization, multi-skilling of workers, team working at the shop-floor level and projects at higher organizational levels were introduced. The first teams, called cells, were developed in the early 1980s in the engineering industry. In the clothing industry, team working started relatively early, in the latter part of the 1980s. Today, "process organization" and "process management" are widely seen as the models for good organization and management in the development of Finnish industry (Alasoini 1999). Designing or engineering cultures have not been as important in Finnish companies as in some other countries (Kunda 1992; Casey 1995; Hochschild 1997; Sennett 1998; Alasoini 1999: 15–16). Finnish expertise has traditionally relied strongly on knowledge of production and technology; development from mass production to flexible production has proceeded from this perspective. Cultural change has sometimes followed, sometimes painfully lagged behind technological and organizational changes (Kortteinen 1992; Niemelä 1996; Lavikka 1997; Kevätsalo 1999).

In the companies studied I met people who had experienced successive crises at work: several company bankruptcies, company mergers and dissolutions. Sometimes they were performing their tasks in the same factory hall as before, but they had seen a constant flow of changing owners, managers and development policies. The work had changed, too. The content of their job might have grown; there were new numerical control (NC) machines and ICTs to be learned, there was a new customer-oriented style of working in a team, there were new demands for commitment and a result orientation, but on the whole the basic characteristics of their life-world had not changed. They were still in the insecure position of a hired worker who was "supposed to work under the leadership and control of an employer" as long as the employer was able to deploy their labour. The most crucial change was connected to what was referred to only vaguely as an attitude, an atmosphere, a different feeling. People have had to change somehow, to be able to cope with the constantly changing, more insecure environment. It is this changed disposition, this new sense of self, that I trace here.

In only two decades, development in the Tampere region has come a long way. The region was the industrial centre of Finland, with redbrick mills of the textiles, clothing, shoe and leather industries, along with paper and pulp and engineering industries. The city was known as the red city of Finland, with active unions and political domination by labour parties. After industrial restructuring, production activities began to wind down, and new activities, such as schools, museums, conference halls and media industries, moved into the former factory buildings. The industrial institutions that had employed people for generations and formed the basic structure for their living disappeared in a few years. Prosperity turned into mass unemployment. Politically, the labour majority on the city council became the minority. The city's former identity broke down, with a multi-level identity crisis of its working-class citizens. What has been for generations a solid basis for life and identity practically disappeared in a few years.

Since the early 1990s crisis, information society development has been very fast in Finland and the Tampere region. It came to replace the traditional legitimating industrial identity (Castells 1997: 8) with a new identity constructed by the international success of Nokia's mobile phones (Nokia was originally a traditional industrial company in the region producing cables, tyres and rubber boots), and other positive aspects that are linked to information society development. The current problems in the information economy have not changed this orientation and the high hopes for the new technology. The city council has launched a new development programme called eTampere, which extends both to industries and research and development (R&D) institutions, aiming to reach a global position in IT progress. However, this development has not brought success for all; the situation of middle-aged and older workers with weaker educational backgrounds is especially difficult. When they have lost their jobs, they have ended up as long-term unemployed. Accelerated social polarization is taking place in the Finnish information society. A gap is opening up between the knowledgeable and the resourceful, who are able to cope with IS development, and those who are at risk of exclusion. However, the picture is not as simple as that; the knowledgeable also have their problems, such as growing pressures and relative "exclusion" from other spheres of life outside their paid work, such as family, hobbies and other social life outside the workplace (Blom *et al.* 2001).

Understanding the informatization of work

New trends at work can be called, depending on one's emphasis and point of view, either informatization (Schienstock 1997) or knowledge intensification (Knights *et al.* 1997). In manufacturing companies, the jobs of upper white-collar workers represent an ideal type of information or knowledge work with dimensions of innovating by means of the demanding and creative use of ICT and

social skills, networking and implementation of innovations. High-quality theoretical, social and technological skills are needed in these jobs. But the
information intensity in shop-floor work has also much increased. Multiple technical skills are a self-evident demand on the shop floor, complemented by good
social skills, motivation to continuous learning, commitment to team performance, and responsibility for the outcome, for the quality of products and
keeping the promised delivery times. Knowledge of ADP use and language skills
are important for shop-floor workers, especially in the units of multinational
companies. Personal characteristics of employees gain in importance as interaction intensity grows along with knowledge intensity. Managers often understand
their tasks as achieving results with the help of people instead of top-down
commanding. In order to gain the profit goals of the company, some managers
need co-operative people. To get the most from the workforce, they need to
support the motivation of their people to become masters in their field.

The means that companies have at their disposal to master varying and
demanding processes include ICT, increasing co-operation and continuous
learning. Knowledge intensification of work is the driving force for the learning
that is intertwined with problem-solving in everyday work practices and also
requires regular extra study. Development of and changes in ICT systems and
software and company organization create much additional work in learning
new practices. The goal of corporate learning can also be a personal transformation process, changing people's attitudes and perspectives on work. The
common interest in the manufacturing case companies was to organize work
community training for employees in order to improve their social abilities and
the quality of their interaction, and support personnel in internalizing the new
collective values in the company.

The new dynamics

The case studies indicate that new horizontal and vertical integration and the
allocation of skills and competences that cross traditional hierarchical, functional
and professional boundaries are being developed in companies for them to
successfully solve problems emerging from the production process. In addition to
the practice of solving acute problems, the emerging co-operation at work across
the boundaries of different communities of practice seems to form a dynamic for
continuous learning and the development of individual and collective competences in the companies.

At the shop-floor level, there are constant time and cost pressures to make
ends meet. Richard Sennett (1998: 57) has written that it is the organization of
time that joins together the other elements of a flexible regime. People struggle
to fulfil customers' needs concerning quality and delivery times with the minimal
possible time and material resources. People need to ask help, advice and information to clear up the complex situations wherever in the organization they may
be found. Besides quality and time competition, there is price competition, as
well. "It is the new management method to promise customers more than is real-

istically possible. It takes the strength out of us," as one of the worker interviewees explained, when talking of increasing pressures felt on the shop floor.

The organization of time is a special characteristic of the new flexible
dynamic of operation in companies. There are many schedules which need to be
met along the business chain from marketing to production and deliveries to
showing growing profits on the bottom line of the quarterly account of the
company. All the individual and team schedules in the companies are orienting
themselves towards some, usually too tight, time limit. Those who have the
power to set these time limits have the ultimate power in companies (see Sennett
1998: 59). Time management disguised as "pleasing the customer" or
"economic necessity" seems to be the central form of control and dominance in
the new industry, which characterizes every individual act throughout the chain
of "autonomous" teams.

The employees' relationship to their work and the work problems they face
has changed. In addition to immediate performance, employees supervise the
production flow over a large area, and are responsible for the target and the
quality of the products together with the other members of their work group.
Their work requires extensive knowledge and a good memory, since the whole is
comprised of many different components. Sensory, implicit or tacit knowledge
related to working on some material with manual tools or mechanic equipment
has been complemented by explicit knowledge, such as knowledge of ADP-
assisted machinery programs and manuals. This is theoretical and conceptual
knowledge, which one has to study and remember to use it successfully in
varying situations. Innovation, be it in the form of a new product, technological
system or work organization, is for employees a new problem to be solved to
ensure a fluent work process. Employees have to develop new work routines and
skills to be able to apply the innovation to their work routines, and guarantee the
undisturbed continuity of the work. Product and process innovations have to be
integrated into the existing system faster than before, without shaking the fluent
flow of the production process or jeopardizing deliveries.

One can see signs of the fact that a certain kind of division of labour, a new
kind of specialization, is entering working life once again: upper white-collar
workers specialize in producing innovations, lower white-collar workers
specialize in co-operation and integration, and blue-collar workers specialize in
securing the flow of the production process and securing the quality of the products and delivery times (see Hirschhorn 1988). Specialization seems to take place
between different personnel groups often within co-operative and interactive
networks, which also form a structure for continuous learning from each other.
This process calls for good social skills among all the participants of the business
and production chain.

The demanding work process also demands a lot from and takes the strength
out of workers. It leaves little or no space for other activities outside of work.
The organizational limits of the demands for increasing flexibility of personnel
in companies are starting to become apparent. Increasing pressure and stress,
growing numbers of cases of depression causing sickness leaves, children's and

youngsters' disturbances are examples of the urgent social problems that societies need to solve these days (Lehto and Sutela 1999). Such problems can no longer be treated as individual maladies; they have their roots in working life. One of the worker interviewees, a young man with a working wife and a small child, described his feelings concerning the new demands of work:

> 'Cos a human being can't devote himself only to work ... It sure has changed strategically, that working life dictates family life. It's this rush. It's the result of this pressing society and as big cutbacks as possible, so that people burn out ... It's just a matter of time when it starts to erupt.

The problem of growing pressure and work stealing time from families does not only apply in Finland. Arlie Russell Hochschild has found out that there is a new time bind (work becomes home and home becomes work) in the life of US working families whose life is centred more and more on work, to the extent that all the other aspects of life have become subservient to working and work and family roles have got mixed up (Hochschild 1997). In addition to the demands of flexible work, the individualization of life in Western societies has become intertwined with family roles and relations. The family has become a kind of jigsaw that needs to balance the different timetables of family members (Beck and Beck-Gernsheim 2002: 90–2).

Organic solidarity, anomie and the transformation of a social order

During the modern industrial era, work has been the primary basis of social organization and the primary constituent of self-formation. One of the many far-reaching implications of a post-industrial condition is the dissolution of traditional and modern bonds of social solidarity and the metamorphosis of the character of the modern self. The shift in industrial work towards forms of production that are discursive, symbolic and highly integrated alters relations among employees and with the physical environment, and has effects on self-formation (Casey 1995). Mechanical assembly-line in production was the symbol of modern industrial culture, and Taylorism its programme of social change, way of government and administration, even ethic (Virkkunen 1990: 41). Changes in the logic of production and society from Taylorism to flexibility have created a new powerful symbol: the co-operative team based on individual and collective reflexivity of post-industrial workers.

For Durkheim, societies pass from a state of mechanical to organic solidarity, a process which is necessarily determined by the division of labour, whose function is to create in people feelings of solidarity. Mechanical solidarity, which characterized pre-industrial society, is defined as a structure of resemblances linking the individual directly and harmoniously with society to the point that individual action is spontaneous, unreflective and collective. In contrast, the basis of organic solidarity is the division of labour and social differentiation. The social structure is characterized by a high level of interdependence, industrial development and moral and material density. Solidarity through social likeness is replaced by that through difference. The individual is no longer wholly enveloped by the collective conscience but develops greater individuality. In this situation, it is necessary that

> the collective leave open a part of the individual conscience in order that special functions may be established there, functions that it cannot regulate. The more this region is extended, the stronger the cohesion which results from this solidarity ... each one depends much more strictly on society as labour is more divided; and, on the other hand the activity of each is as much more personal as it is more specialized.
>
> (Durkheim 1964: 112)

Durkheim's thesis on development from traditional society to industrialized society characterized by organic solidarity does not mean a total break in the type of solidarity. Rather, the result is a society that is bound, on one hand, by mechanical solidarity and, on the other, organic solidarity. Organic solidarity becomes a new type of solidarity born inside mechanical solidarity, its range of development depending on the extent to which division of work and type of market exchange fit together (Kortteinen 1992).

My interpretation of the new co-operative dynamics, alongside the search for market flexibility in manufacturing, is based on the idea of organic solidarity developing inside traditional companies now orienting themselves to the information society and global markets. However, even if the form of social solidarity is changing towards organicism, the sources of anomie and inequality remain. It seems there are two sets of norms and values in the companies: one for the workers requiring endless commitment and responsibility, another for the face-less company owners largely freed from their former obligations towards employees and local communities hosting their production units.

The self as a project

The focus of my analysis is the relationship between information-intensive practices in post-industrial production and the formation of the self: the experiencing and acting self that is able to give meaning to changing work experiences in manufacturing companies. The dominant themes in modern literature on the self come from social psychology, psychoanalytic thought, symbolic interactionism and empirical psychology. In Freudian theory, the self is shaped by the way in which an individual deals with instinctual drives and the associated anxieties from obstructed satisfaction of libidinal desires (Audi 1995: 285–7). The psychoanalyst, Veikko Tähkä (1996), defines the self

> as an actively experiencing and acting organization achieved by a mind, an organization which allows an individual to experience herself as existing

and observe the separate outer world on the basis of her experiences and to interact with it. The self is thus a continuously extending, differentiating, and experiencing subject who has images of herself and of objects that are of varying exactness and duration, and as well as functions which are experienced as her own.

(Tähkä 1996: 91–2)

Sociologists and social psychologists since George H. Mead have stressed the process of interaction in forming the self. For him, the precondition for the self was membership of society; the self-concept arises in social interaction as an outgrowth of individuals' concern with how others react to them. Different aspects of I (= response to attitudes) / me (= social situation in an individual's experience) could be developed by adapting oneself to certain environments through routines, responsibilities and opportunities for thinking (Mead 1976). His work was further developed by Goffman (1963, 1974). Using the dramaturgical metaphor, he focused on face-to-face interaction in situated activities, experiences of interacting individual actors and use of roles.

Many current writers in the social sciences prefer the concept of identity or subjectivity to that of the self. Stuart Hall (1999) has analysed "the birth and death of a Cartesian subject", tracing the version of the modern human subject that had certain fixed characteristics, and a stable understanding of one's own identity and place in the world order. However, he argues that this identity has now completely fragmented. French philosophers, critics and psychoanalysts, such as Foucault, Derrida, Lacan and Kristeva, have encouraged a departure from the concept of "self" in its modern humanist meaning as the core of one's being. It has been challenged as being a too fixed and solid entity (Kristeva 1995) for representing the broken, fragmented lives of contemporary people (Bauman 1995). However, Charles Taylor (1989) has argued that the concept of the "self" has always been more contingent than these critics suggest. For Giddens (1991), in contemporary life challenges the quest for the self is vital for people's everyday working life where reflexivity is needed. Within late modernity, the self is a reflexive project for which the individual is responsible.

My own methodological point of departure on the self is Heidegger's phenomenology and concept of *Dasein*. It is our everydayness, our ordinary, prereflective agency, in the midst of practical affairs, that makes it possible to understand being-in-the-world. We are thrown into the world, which opens to us and to which we as human beings have a relationship, and are given the possibility of finding/not finding our true selves. An individual *Dasein* has either chosen these opportunities or s/he has grown into or got these conditions. It depends on each individual *Dasein* whether the existing opportunities to develop one's true self are used or are neglected. A person, however, is not equal to *Dasein*, but *Dasein* means living, from their point of view, personal freedom and responsibility. Thus a person is responsible for their existence and choices, and is to be understood as an opportunity to become something or someone (Heidegger 2000: 33). The goal is to understand the meanings workers give to phenomena in their life-world at work and their self-strategies created to cope with the changes there (Heidegger 2000: 14–16).

The experiences of people are bound to circumstances and concrete situations, whether repressive or supportive. Taylorism treated a worker as a skilful pair of hands without mental or intellectual faculties. The flexible production paradigm is designed to deploy workers' mental and intellectual capacities to the full for the purpose of serving capital accumulation. It is also seductive as it puts an individual and their abilities on a pedestal and gives them an opportunity to learn and develop. Its demands – workers' rising competence, effectiveness, loyalty and commitment – are limitless. Workers' lives should be one-dimensionally centred on work or serving work-related needs. New flexible capitalism, however, has no intention whatsoever of keeping its implicit promises to workers in return. According to Sennett (1998: 47), the elements forming the system of power of flexibility are the discontinuous reinvention of institutions, flexible specialization of production, and concentration without centralization of power. These all have harsh personal consequences for workers. The new capitalism is not responsible for the employment or fate of its workers, say, after re-engineering or relocating a company's functions. The decisive criteria dictating corporate decisions and actions relate to the best possible circumstances for global capital accumulation. The space for development of the true self has strict limits in this post-industrial production environment. Exploitation and oppression are not history. New oppressive forms entwining workers' selves have been developed, tying people more closely into the search for contemporary business success.

There are good grounds for doubting that the present era is more productive than the recent past. Sharp, disruptive changes in companies lead to decreased motivation and morale of workers, while stock prices often rise with institutional changes:

> short-term returns to stockholders provide a strong incentive to the powers of chaos disguised by "re-engineering". Perfectly viable businesses are gutted or abandoned, capable employees are set adrift rather than rewarded, simply because the organization must prove to the market that it is capable of change.
>
> (Sennett 1998: 51)

In the workplace, Sennett suggests (1998: 55–6),

> the managerial overburdening of small work groups with many diverse tasks is a frequent feature of corporate reorganization. To make such experiments with tens or hundreds of thousands of employees requires immense powers of command. To the economics of inequality the new order thus adds new forms of unequal, arbitrary power within the organization … Control over work groups in flexible organizations can be exercised by setting production or profit targets for a wide variety of groups in the organization which each

unit is free to meet in any way that seems fit. This freedom is, however, specious ... The realities of supply and demand are seldom in sync with these targets; the effort is to push units harder and harder despite those realities, a push which comes from the institution's top management.

In researching US white-collar high-tech knowledge workers, Catherine Casey studied corporate change processes and designed culture: corporate colonization, which offers promises but also obscures contradictions. She found different coping strategies, which she defined as types of selves or self-styles, such as defence, collusion, capitulation. The defensive self is characterized by multiple forms of small-scale resistance. People of this self-type perform their jobs well, protect their everyday personal workspace and go along with the new cultural requirements only minimally. The colluded self is dependent on the company to the point of over-compliance, with compulsive optimism in corporate beliefs. The capitulated self is often a previous defensive self which has resisted the psychic option of collusion, and now is worn down into capitulation. This self-strategy is one of instrumental pragmatism in how to play their game (Casey 1995: 163–78). These findings in a US company have some relevance for what is happening in European, especially Finnish, workplaces and white-collar and blue-collar workers' self-strategies. The self-strategies of a male and a female shop-floor worker and the specific characteristics of Finnish company cultures are now examined in more detail.

Fulfilment or slavery

The focus is on employees' internal mental processes in constructing the self in the emerging knowledge work of traditional manufacturing industry. I have described the framework of the story, consisting of the structural changes at work created by the economic imperative of the flexible information economy, stressing the individual worker's commitment and multiple resources and the co-operative interaction of workers. What I have not yet discussed is how the knowledge intensification of work and a company culture demanding individual responsibility and commitment transform employees' ways of being at work. If new and more interesting work becomes a means for fulfilling workers' own personal needs for development, its positive meaning for employees grows and people may be seduced to a way of life that excludes private living spheres. It may also lead to the intensification of work to the point of exhaustion.

I now turn to the technical, organizational and cultural changes in two companies (a clothing company and an engineering company) and let shop-floor workers, a woman and a man, comment from their own experiences. The research process consisting of interviews and observations in companies was based on focused ethnography (Knoblauf 2001), which both allows one to find answers to research questions and enables a coherent interpretation of research material. The method of analysis is developed from working with the whole data and paying attention to links and contradictions within that whole instead of coding and retrieving the material (Hollway and Jefferson 2000).

The case of the clothing worker: "I feel like a killed worm ..."

The clothing company is one of the biggest and most modern Finnish clothing companies, with a history dating back to the beginning of the twentieth century. It has a good reputation both for its products and for its employment policy (no casual labour, full-time contracts, unionized personnel). Its brand names and trademarks are associated by domestic customers with quality, patriotism and export success. A lot has changed in the company during the approximately twenty years when I have every now and then visited it. Its former production system was a textbook example of Taylorized mass production and long batches well suited to line working. A good pieceworker could achieve quite a nice pay level in an extremely well-organized and polished line-work system using electronic conveyors. The company was forced to give up line working with smaller batches following the change from mass markets to more individual demand from the middle of the 1980s. After experimenting and piloting, teams were established at the end of the 1980s. Other organizational and technological changes followed the change of work organization. The whole business was reorganized in the early 1990s by forming seven independent profit centres organized around their own designs of brand names. Administration and production functions are considered as two service centres offering their services for the profit centres.

Cindy (not her real name), working as a seamstress in the clothing company, is a woman in her fifties. I had met her earlier as a union paper reporter, asking about her experiences in the early steps towards team working in the company. This time we met in a strained situation; the company had informed workers about a major lay-off scheduled for a few months' time, but the names had not yet been announced. We are about the same age: "big sisters" born in big post-war families in the Finnish countryside, who left home for work and for studies, then became working mothers, and finally working grandmothers. We belong to the same generation, and our life course has proceeded in the same rhythm but taken different turns. Cindy has worked for the company from 1969 onwards. At about that time, I started my studies at the university. Cindy's career has proceeded from sewing one straight trouser seam (an operation of a few seconds) to working as a member of a production team and to acquiring the multiple sewing skills needed in teams, and then to sewing entire model garments for marketing purposes in a model-team. To work in a model-team is a highly demanding position at the pinnacle of a seamstress's career. My life has taken me to work as a journalist for eighteen years, then back to the university to do research. Our mutual interest through the years has lain in factory work, the last ten years in team working. Factory work and its transformation is the everyday context of Cindy's work and life; for me, it is a research theme.

I met Cindy in the union representative's small office just round the corner from the sewing hall. We had some privacy, but the hum of the factory was still there. When we met for the first time about ten years ago, there were about 600 seamstresses sewing; now there are eighty. Cindy told me how she now felt about her work:

> It is perhaps this situation … When people are under strain, like us in the hectic piecework, they often cannot take much more. We have had a tough time in the production teams lately. We do not have the strength to take much more: the feeling "this is the last straw" is there. After my shift I have felt like a killed worm. On my way home I have to take many deep sighs to blow out the weariness in me in order not to bring it home. The people at home are not responsible for how I feel … As you know, negotiations are going on about redundancies. We believe that in a month we will be informed about the names of those who are going to be laid off. I suspect that my name is on the list as an over-aged worker. They have not yet told us the numbers, but usually the cuts have been quite heavy … A lot of production has already been moved abroad in search of cheaper workers; they do not need us here any more.

Even if I already knew about the threat of redundancy for the workers, I was slightly surprised about how low Cindy felt. What was happening in the company? During the last eight or nine years it had re-engineered its business and work organization in several waves. News about upcoming redundancies has always been sudden. People have been frightened, but somehow not totally lost their optimism. I also remembered Cindy as a cheerful woman and even enthusiastic about teamwork, with the all new challenges.

Cindy's work includes all the different operations and tasks needed in sewing several types of quality garments belonging to one of the company's brand names. The team does the assembly work on a garment. They get cuttings and special instructions for the garment, including the standard times for its operations, as a bunch of papers, and then they are responsible for the rest. They reserve machines, make sure threads and other accessories are in supply, and generally organize their work. They are responsible also for the quality of the garments and fixing their mistakes. Cindy has also specialized in the technical problems of the several types of machines (about thirteen different ones) used in the team and programming the special embroidering machines. After more than ten years in Taylorized line work and performing one very short operation at a piecework pace, the transition to teamwork was an enormous challenge for her.

> I recall that I had nightmares. I thought that I would never learn all that is needed, but still I learned. It was after only one month when working started to feel like it was normal in a team. I was lucky to have a very skilful person in my team who had worked in model sewing for a long time. She could teach us well.

The social aspects of teamwork are crucial. Cindy talked about the good co-operation in their team, and between the teams. People have also learnt that a friendly attitude towards others makes the work go in a more fluid way in a team:

> We have learned to appreciate others and communicate among us. Clearly our ability and need for communication has grown at work. If you are angry and hostile, it comes back to you sooner or later. If you ask somebody to repair her mistake, you have to do it in a friendly way, not to accuse or to mock. Then you get the work back sooner. A strained person does not stand much.

The change from monotonous line working to teams has had an effect on Cindy's life outside work, too:

> In ten years I have changed my way of life altogether. I have practically turned everything around. And I see that I feel much better now than ten years ago, even if I am ten years older. It is the case that the more meaningful and changing work does not wear one out so much physically and mentally. It was the monotony of line work that used to cause permanent injuries to people. Since starting teamwork I have learned to think more highly of myself, too. I have the strength to exercise, to go dancing and to take care of myself in other ways during my free time. All this helps me also to cope with the pressures at work.

When we talked about team working, Cindy's voice got livelier and I saw a glimpse of her old happy self:

> The work in the team is really varied and interesting. And I must admit that I like it. I like challenges. But then, when I finally started to like our work, it seemed that it was being taken away from us.

Still, there is pressure and strain in teams, too. The company has not given up the stressful piecework system, but has developed a version of it for teams according to which the total number of garments completed by a team is divided among the members as a basis for their pay. The pay system is based on standard times of operations. From Cindy's point of view there is no need for the control that is included in the piecework:

> We are working just as diligently, carefully and effectively on an hourly basis as in piecework. The rhythm of the work is the same. There is no difference. The only difference is that the stress of piecework drops out in an hourly payment system. There is no need to think all the time about how we are doing financially. We think that of course we do our best in hourly work, too, of course we do the work we are paid for. But perhaps those in favour of piecework are thinking that we are tempted to start to slack off, after all.

There are other sources of stress in seamstress work, too, than just piecework. 'Just in time' means sometimes extremely busy situations in teams:

> During this week we have had ten-hour days, because the company is late with the models. And we have to complete them before we can start our summer holidays. It happens that a bunch of cuttings comes to the sewing team, and we have only a couple hours to complete the work. Sometimes the guy is already sitting in his car, waiting for the models from us. We feel a lot of stress then.

ICT-mediated network

Little by little during my fieldwork, the current situation in the company unfolded. The situation was not at all as serious from the company's business point of view as it was from the workers', feeling the threat of unemployment. More production was moved abroad in order to achieve better price competitiveness in the market. The order books had shrunk slightly during the last season. What made Cindy feel so desperate was that she could no longer see any future for herself or other seamstresses in the company, even if they could further improve their working practices. The perspective of saving their jobs, that had encouraged them to strive to be more flexible and acquire more skills, seemed to be gone. The development had now taken a new turn that meant that the seamstress as an industrial occupation was no longer needed very much in the company's Finnish sites. The jobs that are expected to remain are a few seamstresses for the needs of flexible customer service and to assist in design. Already at the time of the research (1998–2000), the majority of the company's personnel were people in white-collar jobs, such as engineers, technicians, designers and marketing people, and development in that direction seemed to continue.

The last ten years in the company have been a period of constant change in its business strategy and downsizing its domestic production. Hundreds of production workers have been made redundant in Finland, while production has been subcontracted to cheap-labour countries, such as Russia, China and the Baltics. The majority of sacked workers have been middle-aged seamstresses, who have become long-term unemployed. They have had very few work opportunities in the labour market, which demands totally different skills than those possessed by the former seamstresses, without formal education or vocational training other than long work experience. Currently, instead of an excellent production unit at its main site, it is ICT that is the number one issue for the company, strategically and operationally. A unique and sophisticatedly integrated ICT system has been developed for the company functions and co-operation with its suppliers. The company's production nowadays is carried out in a network of production units dispersed in many countries on two continents. The company operates in volatile markets with extremely fast-changing and variable production, with a mixed production concept. The network organization is composed of several profit centres forming the company organization, together with a network of suppliers, subcontractors and customers. The key question for the company is how to guarantee the demanded level of certainty in the operation of the network. A well-functioning flow of information is considered to be one of the most important assets.

Good social skills, good knowledge of the company's own business and networked production system, and good ICT skills, in this order, are considered important for their personnel in the company. All people, besides the seamstresses, who use ICT in the company are trained on a continuous basis. However, it also takes individual effort by workers to keep their skills up to date. Competence in ICT is considered to be an individual qualification, which each individual must take care of on his/her own initiative. However, thus far the workers in sewing teams do not need ICT more than marginally. In a teamworking organization, social skills are more crucial in deploying other types of skills, both in the office and on the shop floor.

One way ICT affects seamstresses is through a deck of cards connected to the real-time production management system. Each seamstress has several plastic cards with a magnetic band containing different types of information in the form of a bar code, which one has to remember to feed into the system by pulling the card through an electronic reader. Cindy told me about the functions of these cards:

> Each one of us has a personal electronic card. For instance, in a production team a person must at first pull her personal card, then piecework, then production card and finally a team card. When I go for a lunch break, I must pull my card out, and after lunch break first the personal card and the other cards in the same order again. For hourly pay there is a different card to be pulled; for the inspection of products there is still another card. Together these cards make quite a thick deck.

Cards serve to gather information for the counting of workers' pay, but also of the throughput process of the products (each product has its own bar code) and control and monitor the work process. With this system the company makes sure that the work process is as transparent as possible. On a bad day there is no place for a worker to hide, and if there is a problem in production, it can be quickly traced.

The autonomy of workers in teams is strictly limited to the flexible organizing of assembly work among them, and orienting themselves to the effective reaching of the team target. Departmental organization of the workflow is in the hands of a supervisor, whose job is no longer the immediate and close supervision of workers: they now take care of these tasks themselves. The supervisor also has a PC to feed into and to check information. One of the supervisor's new tasks is to produce, with the help of PC summaries and background information, reports on the achievements of her department for the production management. Cindy's comments on the role of her supervisor stress the importance of mediating the information:

If there is some problem with the model at hand and we need to talk to, for example, the designer, then it is the supervisor whose task it is to take care of this problem. We hardly ever need her in the team nowadays.

The echo of new culture

The management culture and the work climate in the company are traditionally known as "fair". It was the reason why Cindy, too, has liked working there for twenty-nine years now. Cindy stresses the importance of the open and free company climate where hierarchies do not matter in everyday interaction:

> When I started here in 1969, I came from another clothing company. At first I was quite astonished about the open and equal atmosphere here. All the people, including bosses, could be called by their first names by anyone; even the CEO is Perry [not his real name] to everybody ... Anybody can talk freely with each other in the company's lunch cafeteria. There bosses and workers sit side by side, eat the same food, tell jokes and laugh together. It is like all of us are equal and appreciated in the same way. I have liked it here. This is a good company compared to the two others where I was employed before starting here. In the other workplaces the bosses were kinds of half-gods. You felt as if you had to bow and to lift your cap in front of them, even if you hadn't any cap.

Cindy is not blind to the company politics, that people who have served the company well for decades are laid off in a quite ruthless way. After several waves of downsizing company production in Finland, she has got used to the cruel rules of the game, which have been felt in the radically shrinking Finnish clothing industry over the last ten years.

The new "American" company culture has entered the company, with its rhetoric of arguing for flexibility and a committed work orientation. I heard its echo in discussing with Cindy the ideal types of foreman and worker, and how these types have changed.

> The boss should have the courage to make independent decisions. If there is, for example, a tough problem of how to make ends meet in schedules, it should be the boss who takes the responsibility for it. And she should be fair. Well, I am lucky because I actually have not had a real bad boss, ever. As to the ideal worker ... the boss says that a seamstress should be able to stand stress. Then, of course, according to her, a worker has to be diligent and flexible, so flexible that she is able to do ten-hour days whenever asked, and has to be able to perform all the operations, and still do it quickly. A seamstress must not talk. This is the employer's wish ... I think that to some extent we are ideal workers. We are flexible, we are prepared to be flexible in tough situations, such as this one at hand when the models were late. Each one of us has knuckled under. And

we have all been in very good health … in that sense we are good workers, too. And we have good skills and we are diligent … But that we should not talk to each other, that is something that I do not agree with. I feel silly just sitting quietly and sewing, not talking to anybody. Some of us live alone and have nobody to talk to at home. I, at least, cannot bear it if I have to be quiet. And we aren't; we talk all about *The Bold and the Beautiful* [the US television series shown in Finland] and about everything that we have in our hearts.

Cindy seems to have her character intact, so far, despite her long career as a team worker. She knows where to draw the line, what is sensible in the management's demands, from her and her workmates' point of view. Perhaps the growing marginality in the company of her shop-floor position gives her clearer insight into organizational policy and shelters her from giving away her autonomy in the sense of "corrosion of character" (Sennett 1998).

Cindy's female boss, however, in whom the management policy is personified for Cindy, seemed to have adopted the rhetoric of commitment and responsibility of the new culture. Cindy did not agree with her:

It is our boss's opinion that you should put everything in your life behind the work, even your private life. You should eliminate all possible disturbances that might interfere with working. The work is most important. I am positive that this is not going to succeed. Everyone has certain personal matters that sometimes demand being absent from work. But I see that the goal of our boss is to create one hundred per cent commitment among us … It is really a difficult demand for the people who still have small children at home. And the household and the family matter to all people. It is not right to demand that you should devote yourself only to work.

In the course of fieldwork I also talked to Cindy's boss, a working mother of two children in her early thirties. I found out that Cindy's statements of her opinions were valid. The boss of the sewing department seemed to have taken on the flexibility ideology quite literally and uncritically, and also tried to organize her own private life according to its principles. In her free time she was studying management and ICT in evening courses after a full day in the factory. When I wondered about her children's reaction to their mother's absence, she answered cheerfully:

The children learn that it is work that is the most important thing in my life. They get used to it.

I think she was a bit disappointed when I didn't applaud. Instead, I felt sorry for her.

The company's survival game

The tough pressure of competitiveness from global markets can be seen in the company on several fronts: technological, organizational and cultural. Changes in the business environment, such as the demand for flexibility in the new capitalism, are the triggering issue for both the organizational and technical reforms in the company. The company was an early user of ICT and local area networks (LAN): the Internet and electronic data interchange (EDI) had already been used for a long time. Also, the organization of the company underwent several transformations during the 1980s and 1990s, along with changes in the company's business environment. The company is striving to modernize its ten-year-old ICT system in order to develop an integrated system to fit into its international networking. The first (automation), second (isolated systems) and third (integrated system) waves of ADP and ICT development characterize the technical development in the clothing company.

The challenges for the company's integrated ICT system are demanding. ICT is needed in managing the information flows of the constantly changing production. It is the goal of the company to get the information on the logistic chain of its functioning in digital form as early as possible in order to cut costs and to make the flow of complicated information simpler to manage. ICT is used whenever possible. The main production sites in Russia and the Baltic countries are very traditional factories lacking an ICT infrastructure. Traditional communication media (telephone, fax) are used in communicating with them, together with regular face-to-face contacts to co-ordinate and control quality and the timetable of production subcontracted to them. Even if it gets very tiresome for employees in charge of organizing this, the goal of cutting costs is so important that the company has not considered moving its production back to the Finnish partners, which have a well-functioning ICT infrastructure.

Besides flexibility, some rigidity is also needed in the company. In managing the information flows of the constantly changing production, there are strict rules and procedures to control the exactness and the reliability of the information. To avoid chaos, the rules and procedures have to be followed strictly in the same way everywhere where information is fed into the system, including where changes are made or mistakes corrected. There needs to be a hierarchy of access in the system, a system of different codes for different people with different degrees of authority in the company. The system is very centralized; the "Panopticon" has been transformed into electronic form. Still, ICT does not seem to replace face-to-face interaction, communication or meetings in the company. For instance, ICT is used to deliver background information, calls, messages and notes to people participating in these meetings and discussions. In horizontal and vertical teams, people need to have close face-to-face contacts to co-operate and manage the good social climate that is considered a basic precondition for a well-functioning team. This cannot be managed by means of ICT only. The relationship between ICT and culture might even be vice versa in the company: it is the company culture and work cultures that have the effects on the practices of using ICT. Rather than depending on ICT, cultural change in this case depends on the ideology of commitment that is creeping into the company following the new flexible dynamics of operation.

The other side of the survival game is that somebody has to lose. The fact that internal and external means of coping with occupational and market changes have been put into use in the company is related to pursuing only business goals. It is not in the company's interests to avoid exposure to social exclusion or to promote inclusion. It is not the policy of the company to bear any responsibility for the exclusion/inclusion condition of the people found unnecessary for its functioning. What happens to them after redundancy is considered as the responsibility of the welfare state, the labour markets and the workers themselves.

At the end of my discussion with Cindy I asked her to name the biggest change that has had the most effect on workers' everyday life in the company:

> The biggest change is the threat of unemployment. Downsizing has already been going on nearly ten years now. People have been sacked or made temporarily redundant. This is such a … sad thing; it has had its effects on us all. Although I have had the opportunity to stay this long after so many hundreds of people have been laid off, it hurts every time when such a thing happens. It is a terrible source of stress to every one of us, also for those who can stay. It is not easy. When we have worked together for years, and even for decades, it is unbearable to see how some of us have to go. And now it is so hopeless to get a new job. That is … we are now waiting for the news of whose turn it is this time.

The case of the engineering worker: striving for pride and honour

The case is the main production unit of the engineering group in a global corporation, a steel giant. It has undergone a tough period of structural adjustment to the changing business environment during the past fifteen years, followed by successive reductions of the workforce. Its products are advanced product programmes for the drilling, excavation, loading and transportation of rock and minerals. The case company is a world market leader in its branch. It was established as a unit of an old and traditional metal and engineering company dating back to the mid-nineteenth century. The influence of this parent company on the history and collective identity of Tampere has been remarkable. Thus the shock was great when this company was cut into pieces in the early 1990s. A new phase in the case company's development began at the end of 1997. An even bigger globally operating steel company from abroad bought it. Currently the case company acts as one of the main business areas of the new owner. This change, from an independent company with its headquarters in Tampere to one plant among several others with its headquarters abroad, means a different business orientation and change of identity for the people in the company.

During the fieldwork, the company started implementing an integrated ICT system. Organizational changes and improvements to the teamwork system were also made. Technological development and organizational changes were originally separate processes in the company, but in the current integrated phase they started to intertwine with each other. The company has traditionally been technologically very progressive, using the latest ADP and ICT technology both in the office and in production. Also, blue-collar workers are active users of ICT as a part of their everyday work. In addition to using and programming automated production machine centres, blue-collar workers use PCs to feed in information concerning their job tasks, to order components from the automated lift-storage system or from subcontractors, or to look at the drawings of the product on the screen during the installation of machines, to name a few examples.

Organizationally the company has striven to follow the trends of downsizing, outsourcing and team working. The group forms a globally acting network-type of organization with several plants, system suppliers and subcontractors, marketing organizations, and service functions. The work organization in the Tampere plant is based on teamwork. Organization and technology have each formed an approach in search of flexibility, productivity and effectiveness. It is precisely the search for increased productivity and effectiveness in the whole business chain that is the triggering element for the different developments in the company. Changes in markets, in ownership structures and in society in general have their effect on the conditions for profitable business. For instance, the new owner decided on remarkably bigger profit targets for the company than the company had gained before the merger. The ability to be responsive to all the different changes is the crucial thing in the company strategy. In this context, ICT projects are considered to be business projects driven by the business goals of the company. The basic line here, as many interviewees stressed, is that when ICTs cost a lot, they need to bring clear business benefits.

Outsourcing began in the company as early as the mid-1980s when the organization was split into profit centres. All its marginal functions, such as cleaning and maintenance, and the preparation of manual tools, nuts and bolts, were outsourced to subcontractors. During the 1990s, this development continued even further. Nowadays the company has many system suppliers as its partners who are specialized in producing certain components of the final product. Cabins, motors and electronic components of the drilling machines, for example, come from the specialized system suppliers. The company wants to concentrate on the key know-how in the production of the "world's best" drilling equipment of special steel. The assembly work, testing and finishing of the machines is also performed in the company.

The company is a good example of how cultural change in a traditional company is much slower than technical and organizational changes. The realization of the rhetoric of commitment is still seen as quite superficial from the shop-floor perspective and the rhetoric is forced to struggle with the remains of the traditional work culture of working-class resistance. Social skills and technical multi-skilling are stressed as highly valued in the company, where the improvement of co-operation and interaction both inside and outside the company are seen as crucial. A globally acting company has high demands for foreign language skills. Even production workers need to master some English.

Walter (not his real name), an engineering worker in his late forties, is a member of the local branch of the metal workers' union that used to be my employer about thirty years ago, while I was studying. Even if we possibly had met sometime before, we had never talked with each other at length before the interview. Walter belongs to the generation of unionized workers who are not so interested in the traditional class struggle any more, but rather in striving for equality and citizenship in the company and in interaction with superiors. Still, he is a union man. The material things in life are important for him; he is used to measuring everything at work with a material yardstick, which is the legitimate core of the "union language". He is a skilled craftsman and also a workers' representative in his assembly department, where machine parts from departments, different system suppliers and subcontractors are collected and get installed in complete machines. He has specialized in the installation of hydraulics.

The company has had a central meaning for Walter as a source of living and identity. Many members of his family have served the company before him; his grandfather, father, godfather and godmother have worked here. His brothers and sisters are employed in the company, too. He started to work in the factory as a young man after finishing vocational school in 1969. Thus his history and roots are deeply intertwined with the company, which has been his one and only employer so far. In a way, he and his family represent a traditional way of life in the old industrial city of Tampere. Walter has lived through the radical changes in the company and watched the continuous march-past of a series of new managers transiting the company, which has characterized the company's management style during the last few years. It has been Walter and other skilled craftsmen like him who have represented the continuity in the company.

When we met in the sterile-looking negotiation room of the factory office, Walter marched in dressed in a blue worker's jacket. He was angry and frustrated. He came to the interview from a meeting with the employer's representatives that had the aim of resolving a sharp conflict going on on the shop floor concerning new weekend shifts. Neither of the parties, however, had bent to a compromise.

A recipe to spoil the team spirit

I asked Walter about his views of team working, management culture and new technology. He did not need much encouragement, but started to speak as if letting off some extra steam from his engine. He told me a story of his experience of team working:

> Actually, it had already started before this current campaign for team working; it was not the effect of this teamwork project at all. The lads of the

group just started to take responsibility for all the tasks by themselves. I think that now these bosses here dream of something like what we already had at that time. They hope to make the work go in the way we already had it going then, without any high-profile campaigns. We had a damned good spirit in doing things, which started under our former foreman. His dealing with people was of such a high quality that I used to think many times of how he could be so smart and cunning that he took care of his job in such a fine and inconspicuous way that we actually started to take more and more responsibility. He didn't compel us in any way. His way of seeing to things was so frank and fair that he didn't need to argue or point out to us how the work should go. It was as if it had started on its own ... In the end we had a very autonomous way of doing things, and nobody was compelling us to act that way ... For example, there is a note pasted on the bottom of the machine where the delivery date is to be written. One of the lads looked at the date and said to me that if we did all the overtime hours that were possible, there was a small chance of getting that machine ready in time ... I have never been eager to do overtime, but then ... there was the commitment to the group's work ... and we did the job.

Walter stressed many times during the interview the importance of workers' ownership of the new way of working. Even if there is a foreman, his role must fit into the work culture of self-confident and proud metal workers (Kortteinen 1992). They needed to save face and keep their masculine craftsman's honour at work. The main mistake of management seemed to be in "compelling" the lads to act in a certain way. The story continues:

Today nobody cares about the delivery dates any more; the machine gets sent to the customer when it is ready to go.

How was the good team spirit spoiled, then, I asked.

It was the way in which these high-profile teamwork projects have been carried out in this organization on many levels ... I remember in the early 1990s when there started to be talk of the so-called soft values and quality in working life, and the top people in the company pretended we were mates with each other. It was easy to notice, at least I noticed, that and these lectures about co-operation and trust ... they were nothing but sort of formal and compulsory acts. It was not what they really thought, they just had to act as if they were serious about these things ... And then, when the ways of operating were being changed in the company, they brought in such new bosses here who did not care at all about these soft and human values, and positive attitudes ... All positive lessons were forgotten, and the bosses started to act again as their old selves. Already at the start their attitude was false and unwilling, and now they began to manage things in such a way that the difficulties here do not end ... Recently, however, I have noticed some small green shoots here and there, that perhaps little by little we could start to patch the system up again. But it is extremely difficult when nothing goes naturally but instead there are jabs all the way.

I asked Walter to give some examples of the "jabbing" referred to, and he began:

> I wonder how many work groups and committees I have actually been a member of during these years in the company. But I can tell you about one of these experiences. Our new production manager was leading a work group planning one of the current production development projects, and he told us to send to that group people to represent the shop-floor workers who would speak out in the meetings and not sit there in silence. We were told that this was the place to state our opinions and to have some influence on the decisions taken. We chose good people to go there, who know about matters and are able to present their views in a sensible way. I know that when these guys say something it is certainly the real thing they say and no nonsense ... They became totally fed up with this work group.

Why was that? In his answer Walter gives a less than favourable portrait of the new managers in charge of development and company culture:

> It was true that we could speak out there. It was, however, of no use, because the people who were in the decisive positions in the company were practically never present in the meetings, or they came late and left early, their mobiles ringing all the time and disturbing the others and the meeting. It was difficult to have a sensible discussion with them. They could not concentrate because they were so busy. Or they were not interested ... In the end they made their decisions without any traces of the work that was done in the planning group ... The group was just window-dressing for the real decision-making system. Workers were invited there to give them the impression that their opinions mattered ... In other words, this is the jabbing I mentioned ... The guys participating in that group then thought, what the hell, there is no point in wasting energy and thought on this kind of work group.

I really enjoyed listening to Walter. He was sharp and clever, with a very clear view of the problems of the company. The decentralization of work and production does not mean decentralization of power in the company. It has been the company policy to declare autonomy, responsibility and the commitment of workers as rhetorical acts but, in practical reality, the control and centralization of decision-making are still very much alive. This is connected to the traditional company culture, which has emphasized different formal positions and the different status of blue- and white-collar workers. Perhaps the very reason why the transitory young men acting as managers are not able to listen to Walter and other experienced workers lies in this culture. Or are they afraid of losing face and authority in front of their subordinates? Or is it really the case that contemporary capitalism is no longer seriously interested in the development of its basic production functions and in its workers? Or is it just incompetent management?

Walter's way of thinking still follows the cultural codes of workers' work culture with the strong ethos of masculine pride and honour. The core of that culture is no longer mere resistance, but there is a demand for the recognition of workers' abilities to really manage and take responsibility for their own work, to plan and organize it in the best possible ways for the benefit of all parties. Sennett has outlined answers to these questions, which seem to arise in large companies all over the world. According to him (1998: 47), the system of power in modern forms of flexibility consists of three elements: discontinuous reinvention of institutions; flexible specialization of production; concentration without centralization of power. He argues that there are good grounds for doubt that the present era is more productive than the recent past. The practices of sharp, disruptive changes in contemporary companies lead to decreased motivation of workers and a lowering of their morale, but at the same time the stock prices often rise because of organizational and institutional changes.

> (T)he short-term returns to stockholders provide a strong incentive to the powers of chaos disguised by that seemingly assuring word "re-engineering". Perfectly viable businesses are gutted or abandoned, capable employees are set adrift rather than rewarded, simply because the organization must prove to the market that it is capable of change.
>
> (p. 51)

He continues:

> managerial overburdening of small work groups with many diverse tasks is a frequent feature of corporate reorganization. To make such experiments with tens or hundreds of thousands of employees requires immense powers of command. The new order thus adds to the economics of inequality new forms of unequal, arbitrary power within the organization. Control over work groups in flexible organization can be exercised by setting production or profit targets for a wide variety of groups in the organization which each unit is free to meet in any way that seems fit. This freedom is, however, specious. It is rare for flexible organizations to set easily met goals; usually the units are pressed to produce or to earn far more than lies within their immediate capabilities. The realities of supply and demand are seldom in sync with these targets; the effort is to push units harder and harder despite those realities, a push which comes from the institution's top management.
>
> (pp. 55–6)

What Sennett writes seems to hold true in the case company and in its management. However, what is different from the US experience of what he describes is that the case company has not succeeded in planting the designed culture in the workers' minds and their way of being at work. Walter's way of thinking still follows the cultural codes of workers' own work culture, with the strong ethos of masculine pride and honour. The core of that culture, however, is no longer mere resistance, but there is a demand for the recognition of the workers' abilities to really manage and to take responsibility for their own work, to plan and to organize it in the best possible ways for the benefit of all the parties concerned.

Good and bad management

Walter, with his long experience and his shop-floor perspective, could offer some good advice to young managers about "human resource management":

> There in the top positions of this production group there are people who have never been able to co-operate with people. They lack a kind of natural ability to be with people ... That is ... if an engineer who acts as our boss – he has told me this – thinks of how to gain some authority on the shop floor, he thinks of the wrong things. I said to him, if you think in that way, you'll never have the authority you want. You're thinking of some tricks to be performed in order to gain authority. A person's authority – it comes naturally, if it is to come. Don't think of performing ... Just be yourself. You have to see that it is not you who does the job, but it is your group that is doing it. If the company's top managers come to pat you on the back, it is because your group's workers have done their job well.

Walter wants to have bosses who have some integrity of character, people who are serious about what they are doing and are able to relate to people with whom they are working through their mutual target. The problem here is that the employees in middle management do not have the understanding of the work culture of shop-floor workers and its presumed "man-to-man" interaction. The role of foreman has changed in the company in both positive and negative ways from Walter's perspective. One of the foremen said that during one day he has about five minutes for each worker in his group; the foremen have been loaded with ICT-mediated office tasks instead of working with people who are supposed to work autonomously. I asked Walter's opinion about the new role of foremen.

> We do not need anybody to stand behind us and give orders. We know what to do, when and how. If there is a problem, we can sort it out by ourselves. Then if a foreman has only five minutes ... there is no need to come and spend even that time with us. But when a worker really has something on his mind and wants to have his foreman's attention, he should have it then. He

should get some response from his foreman. He should get some straight answers. When something has been agreed with a foreman, then it should stay as agreed.

Walter explains that foremanship practices in the company are not always like that. "It makes people feel really silly, when you always have to have a paper signed about every single thing, be it your pay or holidays or whatever."

During the fieldwork, the ownership of the company was changed into the hands of a big internationally operating concern. The situation was new for Walter, too, who during his twenty-nine years in the company had come to know all the managers personally and was able to relate to them. Managers in the company used to have faces and names, they were personalities who were known on the shop floor. He knew them at least by face when they were visiting the production department.

> Those new managers ... In the last ten years we have had seven managers in succession in this plant. The minimum demand for the new people at the top is that they should visit the production unit sometimes, come to see us workers and talk and listen to us. It is bad thing if information does not go directly from workers to top of the company.

From Walter's point of view, the realities of life have been like that. During the last few years shop-floor workers have represented continuity in the company, the changing managers have been only visitors. The new managers' striving to make their own mark on the company's development has led to strained and tiresome development projects. However, Walter has the experience that the top managers and workers understand each other and find common ground when able to communicate; it is the middle management which twists and turns things. The new ownership is going to make a difference in that practice; a totally new managerial type is emerging compared to those already experienced by the shop-floor workers in the company. The interests of faceless owners of a stock company are now coupled with managerial power which is almost as faceless. The doubling of the profit target for the company-after-company merger was the first sign of what was to follow.

ICTs, mobile occupational boundaries, and hierarchies

The company is a pioneer in the field of deploying ICT. It has invested in high quality IT, including the central computing unit, individual PCs, servers, LAN, EDI, Internet, video conferencing and a corporate network. The company is also a pioneer in using NC machinery and ADP-programmed machine tool centres (robots) in its production from the 1970s onward. The latest development phase is the implementation of the new integrated ICT system that was put into use in 1999. It integrates all the information and communication flows into two main parts (administration and production) and uses programs which flexibly communicate with each other.

On the shop floor the different functions that formerly were performed by a specialized employee now can be added as an extra task to the duties of each worker. For example, the ICT system is used by production workers to feed in all the information (product types, hours, vacations, sick leave) concerning their work-related payment. Each worker has a personal code, team code and codes for the different products in process. ICT is an important tool for the outsourcing functions of the company, too. There is a need to manage the short delivery times of the products, and it entails that information on the orders is delivered as early as possible to all the relevant system suppliers. ICT is of crucial importance here. However, in the routine subcontracting, a system of "filling the empty boxes" is followed.

The company culture's effect on the implementation of ICTs shows the difference in how office people and shop-floor people were trained to use the new integrated ICT system. The office people had theoretical training in the classroom, while the blue-collar workers had to practise on-the-job learning in the factory by standing as a group around their foreman, who showed them the most important new commands and codes. ICT has already been an everyday tool in the production for years, and workers have also needed to master the special characteristics of the new system in order to be able to do their work properly. The training practice made them feel that their work was undervalued and they themselves were considered second-class citizens in the work community. Because of the neglect of the workers' training, the transition period to the ICT system at the shop-floor level was quite difficult, with workers' weak motivation to start deploying the new system; mistakes were made and new problems arose that nobody had expected.

The work itself has changed from the former manual performing of tasks. Walter describes the change in the shop-floor work as transformation and gives an example:

> For instance, if a machinist of a numerically programmed work centre should perform something with the old-style manual revolver lathe [a tool that was commonly used in the 1970s], he would not understand at all how the work should go. Nowadays they programme the machine, shut the hood, type the values and it is the machine that does the job. Formerly it was physically very demanding, because it was piecework. A worker had to turn the handle to take a chip away from the piece being worked on, then had to look and to measure and perhaps take off still another little chip. When you worked eight-hour days with a manual lathe, you knew you really had worked at the end of the day.

In assembly work the ICTs do not change the work itself, but they help with information and communication.

We work with the electronically programmed lift-storage robot, which is managed by a PC; we take care of the contacts with subcontractors and the order of the routine parts by e-mail, and we are able to look at the drawings of a certain machine on the screen to check the individual changes to standard variations. In the future we'll be able to print out the drawings, too, if needed. And we can contact the designer about the changes he has made.

Actually the demands for shop-floor work have grown so much that they do not differ much from the demands of a technician's or an engineer's work. Walter told me that

Often the designers come to see us, ask our opinion about some details. The co-operation between us has clearly grown.

The boundary between white-collar and blue-collar occupations has become blurred in practice, but cultural divisions such as occupational and hierarchical borders continue to exist in the company as zombies, the "dead but living" (Beck and Beck-Gernsheim 2002).

There should be a certain kind of meaningfulness in your work. Many of the former boundaries have already fallen that were still here in the 1970s … What I really would like is for us to have the power to plan and decide how we do our job in the team. And if the team then says that the job takes about three weeks, the marketing people and the others should believe it, and not promise to deliver the machine earlier. It is the way I understand commitment. If we commit to doing the job in that time, then it is our business and nobody else's.

On defensive, collusive and capitulated selves

The personnel in workplaces consist of white-collar and blue-collar workers, of several personnel groups, and of women and men. Besides the company culture, there are differing work cultures in these groups, as well as divisions of work and hierarchical orders among the people in a workplace. It seems that organizational changes, the implementation of ICTs and attempts to plant a new designed company culture are reacted to by different groups and different work cultures in different ways. What is true of white-collar workers does not necessarily fit blue-collar workers in the same way. What happens in US companies does not repeat itself in exactly the same way in European or in Finnish companies. Even if the US recipes for designed company cultures have had an effect on management policies in the Finnish companies, too, the self-processes of people working for these companies have taken different turns depending on the different cultural contexts and resources they have in their use. Still there is a common denominator; the general pressure from the global markets that is felt in companies all over the world.

These case histories represent blue-collar workers' work cultures in traditional Finnish manufacturing companies, companies which have experienced a lot of pressures but have survived by turning to technological and organizational development in their business along the lines of flexible specialization and implementation of modern ICTs. In the literature (e.g. Casey 1995; Ezzy 2001) these lines of operation are dealt with as characteristic of a new designed or engineered corporate culture. It is argued that the language, norms and values of these cultures become internalized and dominate employees' subjectivity. The core characteristic of the two blue-collar workers' interviews discussed above is that neither Cindy nor Walter actually show strong features of internalized corporate values or the corrosion of character outlined by Sennett (1998), Casey (1995) and Hochschild (1997) based on their findings on US white-collar workers in high-tech companies. Still, Cindy and Walter have developed reflexivity; they are able to analyse their position at work and work out their personal standpoints. In contrast, the representatives of white-collar employees, such as Cindy's boss, showed more signs of these symptoms the writers above have pointed out.

Cindy's interview reflects a responsible, caring work culture of women workers in teams with cultural codes derived rather from women's traditional ways of being and working together in the private sphere of life than from the company's specially designed culture adopted in order to get people to commit to the company (Lavikka 1997). Cindy knows where she stands; she is able to get both the positive and negative sides of company's policy. Even if her story has a basic melancholy tone, she is able to see the situation as it is. In Cindy's story she tells about her positive change towards a more active way of life and her rising self-confidence and self-appreciation after she started working in a team. For people working in an extremely monotonous line of work, as Cindy earlier did, the transition to team working which demands the use of various manual and intellectual skills represents a major positive change, with positive effects on their own selves, too. It was as if Cindy had awoken to the fact that she can make changes to her life just as she had changed her way of working. She came to find that there was meaningfulness in her new work and she was encouraged to conclude that the same could be said of her whole life, too. Cindy was clearly empowered by the transformation of her work. But the story has a sad ending: she felt threatened after all. Despite all the development she and her mates had made happen or were prepared to carry out, they were about to lose their jobs because they were no longer needed in the competitive game of the company. There were cheaper seamstresses to be hired in other countries to work for the company with the help of modern ICTs. It seems that the new capitalism does not offer any future perspectives for people like Cindy.

Walter's story reflects, on the one hand, traditional working-class resistance and the masculine shop-floor workers' work culture with, on the other, a strong craftsman's pride and an ethos of honour in the pressures of new capitalism (Kortteinen 1992). His speech creates a strong contradiction between shop-floor workers' integrity and managers' false rhetoric, which is expressed while striving for authority over workers based on wrong premises. Instead of taking in the rhetoric on commitment and trust that was offered by the company managers, Walter wants to have true equality and citizenship in the interaction which the new discursive dynamic of the company's operation intensified. He aims at the real recognition of his and other shop-floor workers' ideas, and in general of workers' ability to have viable ideas for the development of the production. In the recognition he seeks is included a real autonomy for workers' teams to decide the schedules for their work It also includes man-to-man-type equal interaction between subordinates and superiors. Only in these conditions might Walter consent to commit to his team and to his work for the company. Until these reforms are realized, "the machines leave the factory to be delivered to the customer – not necessarily on the planned delivery date – but when they are to ready to be sent". Walter's attitude shows features of self-determination, "internalized democracy", that may characterize individuals in the second modern society (Beck and Beck-Gernsheim 2002).

In general, Walter's options are more optimistic than Cindy's. There are prospects for him to seize a position as a knowledge worker. The position of a highly skilled craftsman, complemented with skills in planning and co-ordination, forms a powerful combination which the employers are going to need in the future. Even if his former career prospects of moving into the hierarchy as a foreman seem to be impossible nowadays, when foremen are not needed much any more, the route of developing new skills and seizing a new appreciated position as a knowledge worker is open. However, there is the permanent threat that the chaotic and turbulent markets and their effects on the company's policy might create sudden catastrophic changes to his life. The position of a knowledge worker is not much more protected than that of a former line worker.

There is also the third story involved in the analysis, and it is mine. Twelve years ago I returned to university after eighteen years as a reporter to work as a researcher at the Work Research Centre of the University of Tampere. For the university these latter years have been a period of increased quantitative effectiveness encouraged by all kinds of performance measures. These measures now are a key means of distributing financial resources between universities, faculties and departments. Research work has itself changed from developing expertise to showing expertise by numbers indicating results.

I cannot recall that, during the first years after my return from the union's hectic tempo and constant deadlines to university research work, I had any inner resistance to that policy. The practical view of getting things done in due time and the internalized central meaning of effectiveness were dominant in my thinking at that time. Then I really was not able to understand how important it is to have silent space for peaceful thinking in the middle of everything. Step by step along the increasing experience in research work, my view has changed. Perhaps I have just grown tired of doing research projects one after another with tough time and resource pressures, a conveyor belt of intellectual Taylorism, seldom having time to look back at the larger scenery behind the fast-produced explicit results. Along these years at the university the free space for thinking has got smaller and smaller. The source of my frustration seems to lie in the conflict of academic rhetoric of valuing the achievement of new and important knowledge and the practices of fastening the speed of the intellectual conveyor belt with the demand of numerical performance. Nowadays the conveyor is also speeded up by the competition of international research financing, such as research programmes of the EU. The special feature in my workplace is that there is no need for a designed culture to improve workers' commitment or result orientation. The tough conditions of project research are enough. The rule of thumb is that projects carried out successfully are the precondition for the next project, but even that is not always enough. Good luck is needed, too.

What, then, are my self-strategies in the midst of change in the university? In Finland temporary work contracts are a typical feature of the career of educated women. The university is not an exception. A realistic career prospect for me as a doctor of social sciences is to be employed in temporary project research if I want to work in a job I am qualified for. The projects have organized my life in many ways during the last twelve years. Periods of more or less social life are tied to the demands of the work at hand, and life plans mainly extend to the end of the current project. My consent to these conditions can perhaps be outlined as collusion strategy in Casey's typology.

The reason why the new designed cultures with their "American" rhetoric have met difficulties in taking root in the Finnish companies, in spite of organizational reforms such as team work and the fact that implementation of ICT started early in Finland, lies in the culture of industrial relations in the country. Finland has traditionally had strong unions and a labour movement including both reformist and revolutionist parties. Finnish history includes the painful period of civil war in 1918 when the nation was divided into Reds and Whites. After the civil war, mistrust between the proletariat and the bourgeoisie continued for a long time; it is said that not all the wounds of this war have healed. Recently, after the worldwide collapse of the communist movement, the traditionally strong Finnish Communist Party decided to end its independent operation; its former members joined the "Leftist Union" founded to prevent political bankruptcy of the communists. Both Social Democrats and leftists were, until recently, represented in the Finnish "rainbow" governments (1995–2003) together with bourgeois parties. This national history forms the background of workers' resistance that continues to leave its traces in interactions and relationships in the companies and work cultures. What remains of the proletarian resistance culture is there in the form of a cultural resource and the ability of the workers and their unions to be critical about the tricks of the designed company culture.

This history has also had effects on employers' policies. Fear of labour power was tangible in Finnish political life as late as the long strike by the metal workers' union in spring 1971, when union members voted no to the employers' offer recommended to them by the union. The whole media in the country, besides the small labour papers, united their forces to put the second version of the employers' offer through. It was then accepted, but only by a very small majority. The eventful strike lasted nearly two months and served as a school of resistance for the union members. Walter, my interviewee, who had started his work at the company two years before the strike, participated in the strike and took in the experience of working-class power. I, too, as a young student participated in the solidarity work for strikers, attended their big meetings and marched with them. I felt that power, too, and it later had an effect on my life. It was this experience that guided me after my graduation to serve the labour papers as a professional journalist. For the employers the metal workers' gutsy strike was an experience which on the one hand increased their respect for the unionized workers, but on the other hand made them think about ways to reduce union influence in workplaces.

The Information Society, Work and the Generation of New Forms of Social Exclusion research programme, which compared organizational and technological development in eight European countries, found that the Finnish companies were among the most developed organizationally and technologically. What made this early breakthrough to flexible modes of organizing and early implementation of ICTs possible? The answer lies in the effects of the deep economic crisis of the early 1990s, which affected nearly everybody in some form and forced the Finns to turn to the cultural "unity of the Winter War" (a metaphor describing Finland's defence against the 1940 Russian invasion) in the struggle to save as many jobs as possible. The rhetoric of wartime unity and sacrifice was also present on the political scene at that time, and it was successful, too. The unions understood their most important task as saving jobs, and so they consented and recommended their members to consent to the necessary development in companies.

Casey's typology of self-strategies in coping with new forms of post-industrialized work, the defensive, collusive and capitulated selves, is as a matter of fact a renovated version of strategies of exit, voice and loyalty (Hirschman 1970). For me it is quite easy to put Walter in the defence category, Cindy in capitulation and Cindy's boss in collusion, indicating their self-strategies. I do not, however, feel quite comfortable with this division. These categories seem to simplify self-processes too much, and in a way cut the discussion short. As Casey herself remarks, the strategies are temporary and mobile and change with the course of life. They might also imply different versions and different preconditions concerning white- and blue-collar workers, women and men, people belonging to different generations with different key experiences. I agree with Casey that the policy of globally operating companies follows the same patterns in every country and continent where the companies are represented. And I do not deny the fact that the role of nation-states is diminishing. However, the different cultural background and history in different countries matters in determining the extent to which employees are vulnerable to designed cultures as simulated, artificial social communities, and what cultural resources they have to avoid surrendering to designed corporate cultures. There might even be continuation of the individual self-processes behind Casey's typology.

The boundaries of the flexibility culture begin to show in workplaces and society at large, in the form of growing stress, increasing burn-out, long sick leave, long-term unemployment and poverty. Increasing mental disorders, growing consumption of psychiatric medicines, symptoms indicating children's growing problems, all have the same message: people have increasing difficulties in managing their work and life. The scale of the problems is global; the results of the sharpening division between rich and poor call for global attention. There are even signs that the new corporate self, the survivor at all costs, has started to consider other options. Mere consumerism seems not to be enough to fill the inner void of the winners. The contemporary fragmentary selves (Bauman 1995) without coherent life-narratives (Sennett 2000) are awakening to the experience of betrayal by the master they have served. The survival game has another side: not all are able to always win. The long hours, all the self-discipline and sacrifices end inevitably someday in the experience of disappointment, job loss, sickness or insecurity. What does the former winner have left then?

Beck and Beck-Gersheim's (2002) analysis of institutionalized individualization in the structural transformation of social institutions and the relationship of the individual to society is useful here. Individualization is becoming the social structure of the second modern society itself. Changes in the institution of work have created a situation where an employee has to meet changing situations at work as an individual, but as an individual who works in an integral relationship with other individuals. Family, class, neighbourhood or other collectives can no longer help to solve the problem with standard answers. It is an individual who must decide, who must make choices, who must find a way out of the situation. It is a challenge of reflexivity for an individual, the ethics of which is described as: "Thinking of oneself and living for others. It means giving up and having a lot of dilemmas and paradoxes about it" (pp. 202–13). At the centre of this ethics is the idea of the quality of life, including other things than material success, such as time to oneself, rest, self-determined commitments, relationships and family life.

Note

1 Three projects contributed to this work: Information Society, Work and the Generation of New Forms of Social Exclusion (SOWING) 1998–2000; Mobile Boundaries of the Information Society 1997–2001; and Gendered Organizational Cultures and IT in Manufacturing and Service Industries 1998–2001.

9   In search of boundaries

Changing boundaries in and through teleworking

Riikka Kivimäki

"Work is a state of mind" is how a 22-year-old managing director characterizes his work when pondering the meaning of telework. Time and place do not have importance in his work. This is one of the most important aspects of mobile boundaries in working life in an information society; more and more work can be done supported by information technology. The definition of telework is difficult, not only for this young managing director. Telework can be defined as "electronic homework", "telecommuting" or "flexiwork". These definitions have their own histories. Electronic homework has had negative connotations as unskilled, low-paid office work, either full-time or freelance. Telecommuting is the dominant term in the US and includes people working at home with computers connected to offices many miles away. In Europe the term "flexiwork" refers to the expansion of electronic network accessibility and growing availability of portable computers (Qvortrup 1998: 22–3).

Three people, one woman and two men, tell their stories in this chapter. They do not define themselves as teleworkers. They just do their job and use the possibilities information technology provides in their work. I am interested here in several questions: How they do organize their work? What are the consequences of flexibility in the relationship of working hours and leisure time? Where are the boundaries between the different areas of the totality of life? Do these boundaries exist at all? How does the totality of life change when there are processes that reshape the boundaries of work, such as at a given workplace?

The reality and totality of life is formed locally in historical, cultural and material processes. Reality is multiple and a result of practices. It cannot only be looked at from different points, but it is manipulated by practices. Reality is made and enacted. According to Mol (1999: 75), "if reality is done, if it is historically, culturally and materially located, then it is also multiple. Realities have become multiple." The reality in the case of these three people is done and enacted locally. What are the practices and processes that are shaping the reality of their totality of life? What does the new order of things mean to women, men or the sense of community? What is the meaning of different positions when working life has changed?

Multiple forms of reality

The local and temporal situations of the storytellers in these field stories differ in age, gender, family situation, use of information technology and interaction with employers or customers. The stories tell of the alternatives that the storytellers have had to face in decisions in their working life and of the ways they organize their work. The themes that run through the case studies are "the good life", the sense of community and the degree of freedom in choosing one's working environment. The three cases are: *Managing Director* who owns her firm, who has managed to share her time between her family, work and leisure time and who has made a free choice to do telework; *Game Designer* who is an employee in a firm, who is failing in sharing his time between work and leisure and who has also made a free choice to do telework; and *Freelancer* who had not chosen his position as a freelancer and a teleworker of his own free will.

Telework has not generally been suggested in Finland and other Nordic countries as a way of reconciling work and family life as much as in other European countries. In contrast to Central European countries, Nordic countries provide developed childcare systems. The motives for doing teleworking in Nordic countries differ from those in many other countries. While teleworkers have often been poorly educated female workers elsewhere, in the Nordic countries they are highly educated men and women. Telework has previously been studied in Finland in terms of laws and agreements, labour markets and work organization (Pekkola 1993a, 1993b, 1995; Heikkilä 1995). My aim is to trace emerging mobile boundaries in and through teleworking.

Economic enterprises as well as ordinary people have become more and more interested in the possibilities of teleworking. It has been seen as a way out of unemployment and as a new way of arranging relationships between private life and economic activity. The positive consequences of decentralization, increased worker autonomy and mobility brought about by telework can be seen in higher levels of productivity, improved working-hour arrangements and new employment opportunities for various categories or workers. However, telework can also generate isolation and marginalization, increase stress for workers, and increase their exploitation (Di Martino and Wirth 1990). It is essential that the flexibility of work increases in teleworking in many respects. But many open questions remain. Does teleworking mean flexibility in other areas of the totality of life? With women in Finland working as much and as long hours as men in the labour market, what does this mean, if the work is telework? What does this mean for everyday life in families with teleworking parents? What positive and negative effects could increasing teleworking bring to the relationship of working life and the totality of life (see Huws 1995)? Will increasing possibilities for telework change practices in families?

The positive effects mentioned in Huws's White Paper to the European Commission include the following. Teleworking can facilitate the restructuring of working hours, making it easier to introduce part-time working, job-sharing and other flexible forms of work. It can help to reduce unemployment and bring into the labour market groups previously excluded as they were unable to offer themselves for full-time work. It could improve quality of life by opening up new choices about where to live, through environmental improvements (as an indirect side-effect of lower energy consumption and lower concentration of population in large cities) and making available labour-saving products and devices.

The White Paper also mentions some negative effects of teleworking. They are: the disintegration of collective forms of worker organization, leading to the atomization of the workforce and exclusion from the social dialogue; increasing precariousness of employment leading to economic insecurity and social problems; the exploitation of vulnerable groups of workers such as women with young children, people with disabilities and members of ethnic minority communities; reversal of progress towards equal opportunities brought about as an indirect result of an increasing polarization of the workforce; negative impacts on family life, through continual presence of work in the home; erosion of vocational training structures where these have been traditionally based at the workplace; transfer to peripheral regions of only the most low-skilled and repetitive types of work; and social isolation resulting from home-based working.

According to Minna Salmi's studies in Finland, home-based work changes everyday time structures in terms of the working hours. She found differences between men and women; men follow more traditional timetables than women, who adapt their working hours according to the time structures of other family members (Salmi 1991). The timetables of entrepreneurs also differ according to gender: women's work in enterprises was more connected to family situations than men's (Aho *et al.* 1995; Kivimäki 1996). In Finland, a 1994 study examined how equality between men and women has been developed by teleworkers (Ylöstalo and Kauppinen 1995). According to Heikkilä (1995), 44 per cent of teleworking men thought that equality had increased with teleworking, but only 22 per cent of women agreed. Thirty-eight per cent of women but only 10 per cent of men thought that equality between men and women had decreased.

Working in the home does not break down existing segmented gender dimensions but may even reinforce them (Stanworth 1998: 60). A proper understanding of the link between wage work and the family requires an analysis beyond the interaction between two separate domains. Work and family are present in the activity of real individuals as points of view and values, and as acts and practices. Life is structured as entities, not separate domains. The different everyday lives of women and men are shaped and upheld by the gender contract. The gender contract is connected with labour market structures in which working life and family life, production and reproduction are institutionally separated. But in individual lives these are combined as a totality, and this creates a daily conflict. The relationship between working life and family life varies according to gender. Gender means differences in time use, working hours, childcare leave arrangements, housework, etc., and in hierarchies or the specific features of occupations in the labour market. If work should change to become more independent of time and place, what would this imply for the totality of people's lives?

Typical teleworkers are translators, researchers, data input workers and journalists. The popularity of flexitime has increased in recent years. In Finland employers' attitudes have slowed this development. The high initial costs in teleworking have had the same effect. However, experiences of teleworking have shown that productivity has increased. In telework one can avoid working under stressful conditions and errors may decrease. The vision of improved balance between working life and family life could include that children will be taken care of cheaply while the mother, or father, is teleworking at home. The other, more positive, vision is that teleworking can increase flexibility for all in families. Work can be done at home when conditions demand this. The new technology has at least two faces. New opportunities can be found in new practices. It may be that teleworking for many jobs can help reach a better balance for some working parents and enhance the quality and flexibility of both their working and non-working hours. Or through teleworking "we could indeed be moving into a more flexible future where men, women, employers, childminders and organisers of day nurseries all mesh together into a new creative world of flexible life-styles and rewarding work" (Pahl 1993: 642, cited in Hamblin 1995: 495).

Three stories

Which kind of changing boundaries in and through teleworking can be found? In searching for these new boundaries I chose twenty people for an interview. They are women and men who can decide where and when they work. Information technology is familiar to them and, for most of them, information technology is not only a tool but also an object or product of work. Some of them are living alone, some have a family with children. I found them at workplaces where the work is suitable to do as telework. From twenty interviews I chose three cases in which three themes are discussed: Game Designer, Freelancer and Managing Director. The first theme is the sense of community: what do the teleworking women and men think about the importance of work community? The second theme is free choice versus compulsion to choose telework: Is teleworking the most adequate way to organize work or is it the only prospect? The third theme is the totality of life in telework: how is working life in parallel with private life?

Game designer

A passion for computer games gradually became waged work for Game Designer. He is around 30 years old and is living alone in his city apartment. He can work both in the office of the firm where he is employed or at home. As a sideline he writes articles evaluating computer games for magazines. Just recently he has begun to work more intensive and long periods at home because telecommunications have improved. He designs game spaces for different kinds of products. He is pondering his title or occupation:

I think my title when I was hired to the firm was "programmer", but that is what I do the least. I've shown all kinds of theatrical and expressive know-how but what I'd like to do more is scriptwriting. On my business card my title is "designer". Now I have pressure to change it into something else … If we take an analogy from movies I'd like to be "director". What I am, if we use some American terms, is "project leader" or something like "lead designer". But what I do and what my actual title is, if we give some attribute to it, is "game designer".

While Game Designer is musing over his title, the same thing is happening in many other workplaces where new information technology is shaping structures. Occupations, tasks, hierarchies and titles are in a process of change. Game Designer does not know if he should call his fellow worker his subordinate or adviser:

I may have as a subordinate, or as an adviser, someone who's good at animation. I'm not a graphic designer at all, I can see what's good, but I'm so bad at it that I wouldn't get it done by myself, so I negotiate with him.

Game Designer states that he has not succeeded in separating work and leisure time. He says he can do his job at any time around the clock, so that there is "no sense" in it. He ponders that his life situation can be affected by the fact that he is single; there is nobody at home who could claim some timetables. He can start working at noon, and work in short periods. He can have some leisure time and continue again in the evening into the small hours. The boundaries of working hours and leisure time have disappeared.

Or you can have whole time as free time, so that you do not know which one it is, and that is because you have combined your hobby and work. Because they are so close to each other.

Game Designer's hobby, computer games, has changed in a sense into two different jobs. He writes articles about games, and designs spaces for games. He has nevertheless considered how to separate working life and leisure time.

So, I should get some sense into how to begin and stop working. Or I should get some other hobby to make me leave this game stuff and go to do something else. Well, badminton has been the only activity in which I have noticed that this is something else.

Game Designer does not want to buy a mobile phone because he has a feeling that if he had one he would always be reachable. This could lead to an even tighter commitment to his work, and at the moment that is already too tight. He alternates his workplace, he works both at home and in an office. When the project is in the implementation phase, he works at home as much as possible. In the workplace he does not have a computer or a chair of his own, so there he could not concentrate on his work as much. In the planning phase he has to be in the workplace with his colleagues, because that is when a lot of communication is needed. He is of the opinion that there are some things that cannot be discussed by e-mail in the planning phase. E-mail messages are written, somehow too "ready", and the colleagues take them too seriously.

Now that he has an electronic network at home he has been at his workplace even more. Contacts by e-mail seem to be insufficient in some cases. He says that when he has an e-mail from colleagues, instead of replying he has to go to his workplace and ask what is going on. Work community is important to him in other ways.

> On the other hand I have a desire for the work community, because here [at home] I go crazy when I am here surrounded by four walls.

Game Designer lives near his workplace. Geographical distance is not the idea behind teleworking in his case. What is important is that he can draw away for a while to do the job which demands concentration.

> So it's a priority, when I want to minimize the disturbance, I want to block out those people so that I can do my own part, I escape here, and I have only e-mail contact, I do not have a mobile phone and I may even take the telephone off the hook.

With e-mail he can ration his availability and the point in time when he reads or replies to his mail. Teleworking is an opportunity for him to do his job at home, only about two kilometres away from his workplace.

> It is not telework in the first place but the possibility of getting your own privacy, doing your own part.

In his job, geographical distance would not be good. He said that he sometimes needs to meet his colleagues quickly, to ask what they meant in their messages. His work involves so much co-operation that working at home is only one interim phase. Collecting the elements of a common product demands close physical presence, engaging in genuine co-operation.

Game Designer is in e-mail contact with some of his friends, besides his colleagues, but not with his relatives. He is somewhat bitter when pondering the difficulty in binding working hours and leisure time. In his opinion, it has ruined his private life. "And that is maybe why I'm single," he stated.

Game Designer has come to a critical period concerning the practices that shape the totality of his life. He can now decide when and where to work, how many hours to work or how long some job will take. He reveals that sometimes he has cheated, but it has been unprofitable to him; in other words, he has not reported all the hours he has done.

Hey, now we get to another problem. And that is, when I use 150 hours to do a job for which I'm allowed to use 100, I've done overtime. But my employer does not get money to cover these extra 50 hours. I could then take 50 hours of free time … to somehow clear my conscience. But that's not how it goes. I feel that I've caused a loss to my firm and I should spurt to make up for that too in my job.

It is like contract work, but not in the sense that you could earn more money when you work harder. In the case of Game Designer he has a contract, for he has to compete with colleagues all over the world for the same kind of jobs. He has to do his job so efficiently that he can ensure the competitiveness of his employer and, at the same time, ensure his own future work contract. Overall, Game Designer loves his job:

No, I don't do it only because it is my job. I really have a passion for it. And when it is at its best, it is a passion.

Freelancer

Freelancer is separated and lives alone for most of the year. One part of the year he has his children with him. At the moment he is working as a freelancer, which has not always been the case. The reason why I chose to include the story of a freelance journalist is that free choice versus compulsion to do telework and combining working life and family life are seen from other angles than in the story of Game Designer, who is a wage earner.

In Freelancer's work, as in entrepreneurs' work in general, boundaries of working life and private life are difficult to define (Kivimäki 1996). There are periods in which working is not possible. That happens when Freelancer's children are staying with him at his house. There are too many distractions when the children are playing. He has to get a childminder to take care of his children. But there is plenty to do anyway. He has to cook, wash the dishes, clean the house and wash the laundry. Freelancer noticed the importance of a childminder when he tried to work while his children were at home. He thinks that combining working life and family life in teleworking is difficult:

Yeah, I think that it is nonsense to advertise that teleworking is a new way to combine work and family. It's not true. You can put it in your report, it's nonsense.

Freelancer begins work at eight in the morning. He is used to having some breaks during the day and in the afternoon he works again. Sometimes he has an urgent task to do and working can continue into the night. Then he cannot think of anything else but work-related issues, and that does not feel good.

Work community and sense of community are important to Freelancer. His problem is that he has to work alone so much, although he would like to have contact with people. He co-operates with clients but the co-operation is very businesslike and emotionless, just going through the agenda and nothing else. A sense of community cannot develop as easily as in a steady work contract.

> I think one problem is that I have to do so much alone. It is not good for social contacts. Work community is one social phenomenon and it is rather important. Of course I see these people frequently, but it is clear that in a job like mine there are notably fewer social contacts than in the situation in which I had an eight-to-four job.

At the moment, Freelancer's most important community is his family, his children. He wants to spend as much time as possible with them. He has also in a way made friends with one of his clients and some people around this client who have similar thoughts to Freelancer, although these contacts are mostly by e-mail or phone.

Freelancer said that he is more than willing to work on a steady work contract. His situation resembles "outsourcing". He thinks that now that his work contract is not secure, he has more difficulty concentrating on his work; with each passing day he has to see to it that he has a job in the future. He cannot decide how to do his work: the most important thing is what his clients have in mind. Working as a freelancer is not his free choice. It is compulsory because of various circumstances, and he wishes he could be in another situation. He has written and sent applications to employers. He is pondering how it is more advantageous to an employer to have teleworkers like him.

> Is it so that employers want this kind of teleworking? They may save some expenditures or something.

Working in short periods is not good, in his opinion.

> Maybe the most important difference is that you have no continuous job and have the feeling that your security is missing. I think that freelancers like me – if they are teleworking or not – are in a vortex of insecurity. And it is a rather unpleasant feeling.

On the other hand, there is a feeling of success when he finds a new client.

Freelancer has no choice. He does not like his job. This kind of sense of community is not sufficient for him. Contacts through information technology are not enough to meet his need for social interaction. This is maybe because he has not been able to make his own choice. If he could do so, he would not be working alone. But if he had decided to telework of his own free will, he would probably have created satisfying social contacts and forums for himself already.

Managing director

Managing Director owns one part of her firm. I chose her story because she has experience in working life and business life, but also in combining family, work and entrepreneurship. She has learnt to adapt her everyday life and now she can choose her working hours and workplace. Her work is not bounded by time or place. Routines in her work, if we could call them that, have developed in three places where she works. She can also work when she is travelling between these places. She has created efficient practices in her work, and divided her working days into periods so that all jobs are done. One routine is that she switches off the telephone when she wants to concentrate.

When she is working at home she takes her small children to school in the morning. After that she works actively, sitting by her computer and phones, and she does not do anything else. In the afternoon she fetches her children from school and she is with them until five. She also follows her daily housework routines, which include cooking, doing the laundry and jogging. After her children begin to do their own things she works again until nine in the evening. Her working day is very long and fragmentary, but she has the feeling of having had a whole and productive working day.

> I think that I'm not doing any more work than those who work from nine to five. I can do a lot of work during the day. When I want to concentrate on something I switch off my phones, that's when I want to be left in peace.

Managing Director takes care by e-mail of all the affairs that can be handled that way. When she wants to contact her clients, she first contacts them by e-mail and then calls them. She wants to have more contacts and business by e-mail. She is also in contact with her friends by e-mail. In addition, she has e-mail contacts with her adult children, who do not live with her any more. Competition between family and work has not affected her. She can give an equal share of her time to both work and family. As an entrepreneur this is familiar to her and no one sphere of life is suffering. But this way of working requires self-control.

> It requires a lot of self-control, and that's what I have had to exercise. I am very good at concentrating on something and shutting out everything else. My friends've told me that it'd be possible to blow up the whole house around me. But it is not easy. They think that I'm too tough. But I don't worry about the things I can't help. So I've learned to divide my work into periods. I'm totally pleased with the situation I'm in.

Managing Director uses the Internet for banking services, shopping and information retrieval.

> I'm enchanted by the Internet somehow. I'm enchanted by that whole virtual world, by the possibilities it has.

She mentioned one virtual community in the Internet to which she belongs. Managing Director has built up her work practices herself. She feels that she can control her use of time, for instance by switching off the telephones when she wants to concentrate on her work. Shuttling between three places is her own choice, and that is what she likes to do. She is making use of her freedom to define her working hours and workplace. The Internet and e-mail facilitate work and non-work. Through experience, Managing Director has learnt workmanlike control of time use, which helps her in her work.

Does everyday telework meet the objectives set by policy-makers?

Policy-makers have presented positive future visions of telework, to which the EU and Finland have set out to aspire. What are these visions like when looked at from the standpoint of people who do telework daily?

Game Designer and Managing Director work in a company which produces new elements for the Internet or for a virtual product. Their jobs are the "new occupations" that have emerged along with the new information technology. Freelancer's work has been "outsourced"; in other words, some of the company's tasks are being performed by people with short work contracts. If the company did not have new information technology at its disposal, Freelancer could perhaps still have a steady work contract, which in reality he hopes to have. In his case, the opportunities brought about by information technology may have taken a permanent job away from him. When talking about the increase in the intensiveness of work, interviewees stated that working is indeed becoming more intensive. Employees can concentrate on working intensively when they can determine their working hours and workplaces themselves, and if they can eliminate any factors that distract them in their work.

According to the case studies, reconciliation of working life and family life in telework is varied. It is difficult to reconcile teleworking and a family with small children if work is done only at home, as in the case of Freelancer. What makes working at home possible is a well-functioning childcare system. Threatening visions of an increasingly marginal position of women in the labour market, as reconcilers of short-term work contracts and family, do not seem probable in the case of educated women with full-time jobs, at least not as long as childcare arrangements are provided at the present level.

The better one can adjust one's day to work and other everyday life requirements, the better one can succeed in the reconciliation of work and non-work. If work can also be done outside the home, as in the case of Managing Director, it is easier to draw the line between work and non-work. Failure to do this aroused in Game Designer a feeling that work ruined his private life. "Burn-outs" and negligence in taking care of their friendships or their own well-being, particularly among young men working in the field of information technology, will be increasingly common in the future if they do not know how to manage their time and resources.

Co-operation with customers and other actors varies depending on the companies and entrepreneurs. In Managing Director's work, e-mail and the Internet, which enabled her to do telework, also provided her with an opportunity to be in touch with her clients and other contacts in a flexible way. She hoped to be able to attend to her business even more through e-mail and the Internet. In Freelancer's work, co-operation means several employers between whom he shuttled and from whom he hoped to get additional work opportunities in the future. In Game Designer's work, the intensity of each work phase defined the number of contacts to both his fellow employees and his employers.

Regional redivision of work had become reality in Managing Director's work: she worked in three different places. She chose each place according to the demands set by her work or by her family. Contrary to what she perhaps expected, commuting did not diminish in her case. Game Designer worked at home a few kilometres away from his workplace, because living close to his workplace was necessary due to the nature of his work. Freelancer worked at home, where he occasionally spent time with his children. He had chosen to work in his former place of residence in the country; in other words, an objective set by policy-makers had been attained. He also commuted notably less than if he had been a journalist with a permanent job. One of the threatening visions concerning telework had become real, however. That is the threat of insecurity concerning employment and the financial situation of his family, a threat which he actively tried to remove by seeking a permanent job.

The job description of teleworking people resembles the people who do the job. Their everyday work is a mix of machines, software and matters at hand. E-mail and the Internet provide not only new possibilities, but also limitations to interaction and communication. Telework has been addressed as an issue with one single character, dimension or content. Yet the foundation for telework lies in individual skills, knowledge and know-how, the level of which, in addition to the level of expertise, determines the individual's capability to master the action space and its extensiveness in telework. Telework requires of the individual an established and accepted position in an organization as well as the ability to work independently.

The new electronic forms of the sense of community are possible in the situation in which the job, hobby or interest does not necessarily require face-to-face meetings. Therefore, a virtual sense of community is also possible. On the other hand, e-mail cannot meet the requirements of all levels of interaction, if work requires doing something together, brainstorm-like planning or negotiation. Nor, at the emotional level, does it replace real meetings of people, in which facial expressions, gestures and non-verbal communication play an important role.

In everyday work performed by teleworkers, working manifests itself in a spectrum of practices. Drawing the line between work and non-work requires knowing how to control time use. According to the case studies, this seemed to be easier for Managing Director than for wage-earners, who have to prove their efficiency not only in outputs but also in how they use their time. Do wage earners in their position feel tempted to look more efficient than, according to their time use, they actually are? In other words, tempted not to report all the hours they put in to their employers? The increasing shift of teleworkers from wage work to entrepreneurship brings along with it changes that concern the individual teleworker, the work community and the labour market.

Game Designer, Managing Director and Freelancer each have their own stories of working in an information society which promises a new kind of life. From the standpoint of community, telework creates both threats and new opportunities. There are new practices emerging. One threatening vision includes lone toilers, whose social contacts depend on themselves and who work without the support of a work community, and also without any separation between their work and private lives. When depending on electronic means of communication, the forms of social contacts vary. Short e-mail messages without non-verbal communication make meetings simpler, although some situations require the possibility of meeting face to face. Short meetings with clients or colleagues do not provide space for those social phenomena at workplaces that have been self-evident. Learning on the job, by asking colleagues for advice in difficult situations or sharing knowledge with other workers, may not be easy.

From the standpoint of the totality of life and the "good life" theme, the other two themes, the sense of community and the voluntary/involuntary way of working, are in close interaction with each other. We all probably aim to have a good life. A good life in the work community and in the workplace, in my opinion, depends on the good life of the employees, of which work is only a part. Everyone aims for a life that pleases them according to their personal life situations. Working independently of time or place provides more choice in the reconciliation of work and non-work.

Although telework introduces more ways in which to be flexible in reconciling working life and family life, it can also cause problems. One of these threats is the assumption that it is possible to take care of small children and work at the same time. Another threat is the blurring of the boundaries of work and free time. The concept of overtime is blurred when, instead of measuring working by the hours on the clock, it is measured by work periods during which a certain task is completed. Employees', employers' and work communities' good lives are not necessarily in contradiction with each other. Telework enables employees to work better and better in accordance with their own life situations. This relates closely to the third theme of this article, the voluntary/involuntary way of working.

Organizations whose intention is to increase telework have adopted the idea that it is efficient and productive. There are also limitations to the increase in efficiency, if employees work against their conceptions of a good life which is influenced by other factors and walks of life than the work itself. The change of work contracts into more entrepreneurship-like situations alters other things than the intensity and efficiency of work. These other things, such as the conception of a good life underlying the actions of employees, the totality of their lives, their need for community, health, occupational safety or learning on the job, all require attention to ensure that working will also be productive and effective in the long run.

Where can we look for boundaries in an information society? Should we find them? Institutions do not necessarily strictly determine working hours and workplaces any more. People in the same phases of life, just as people of the same age and sex, as well as those doing the same work, do not necessarily choose the same working hours (Julkunen and Nätti 2000: 201). Networking of work creates new, local possibilities of boundaries that can be defined by individuals themselves (Castells 1997). The mosaic-like formation of time and place in the everyday individual life (Sennett 1998) is a characteristic answer to the changes brought about by the information society.

If it is increasingly not institutions but individuals that set boundaries, how can we devise societal strategies without taking into account the factors that influence families' and individuals' choices: gender, age and community? The totality of life becomes a stage for individual choice on which people aim to have as good a life as possible, scattering it with work, free time and quality time together with family and friends. The concept of working hours loses meaning, and the process of combining work and free time becomes more important than that of fixing boundaries. This novel, process-like living requires unlearning the ways of the old factory-whistle society. Teleworkers have been the forerunners in unlearning this. They have learnt through trial and error to know their individual boundaries which, after all, cannot be set by any others than themselves.

Part IV

Concluding discussion

10 Information society, agency and identity positions

Tuula Heiskanen

Non-gendered and gendered information society

The commonplace understanding of the information society draws on a variety of sources from political documents, public discussion, scientific writings, and practical activities. It is both a scientific concept and a catchword that draws together contemporary trends. Catchwords tend to generalize and universalize. In contrast to this tendency, we have given concrete form to the concept by presenting Finland as a specific kind of information society. Castells and Himanen (2001: 103) have described the distinctive features of Finland as a model of an information society in the following ways: the competitive mobile phone and Internet companies; the state-led system of dynamic institutions advancing technological innovation, creative computer hackerism, imaginative citizen-initiated social hackerism; the combination of the information society and the Nordic welfare state; local information society initiatives; and a national identity that is technology-positive and favours networking.

These elements together have produced a model of society that stresses the integration of institutional objectives related to economy, technology and welfare, even though at a practical level the integration has been and is less straightforward. Contradictions among objectives are visible, for example, in official information society documents. Jari Aro (2001) has illustrated the rhetoric around information society in Finland by analysing official strategy documents and public discussion. In Finland and elsewhere, discourse on the information society is in many ways focused on technology. In those respects, the discussion echoes broader conceptions of technology and ambivalences around it. Official documents, which have relied in their writing on extensive expert work, explicitly reject crude versions of technological determinism which describe technology and economy as independent forces in the social world. Rather, technology is presented as a neutral instrument or a means to reach some end. But the neutrality of technology is also a dubious claim. Aro remarks that there are explicit or implicit requirements that technology needs in order to function. For example, new information and communication technologies carry with them some forms of a technical way of life, such as the need for electricity or the basic skills of reading and writing. In official documents technology is typically seen as both a liberating and a destructive force. The liberating side holds out the promise of wealth and well-being, the destructive side threatens civilization. The beneficial consequences of information society are achieved and the unfavourable results avoided by the appropriate use of technology.

Official documents construct as actors of information society: first, the nation as a whole, and second, individuals, businesses and public administration. What is missing from the documents are the guidelines on how to look at developmental trends from different actors' points of view, even though it is also clear in the documents that the term "information society" means different things to different groups of people. Feminist critiques of technology have shown what the approach from different actors' points of view might mean, and their resulting consequences. One of the most important differences related to information technology has to do with gender. In this book we have thus sought to keep gender in sight.

Marja Vehviläinen (1997; Vehviläinen and Eriksson 1999) has analysed expertise related to information technology. She points to the fact that in Finland women have used information technology in workplaces more than men since the 1980s, and that women have also participated in the planning of information systems: at the beginning of the 1990s about one third of the planners were women. Librarians, most of whom are women, have introduced access to computers and information networks for people outside working hours. However, by the turn of the millennium, women's IT expertise was confined to a limited user-role instead of a broader developer expertise. Also, women's share in the IT professions decreased sharply during the 1990s and home computers have become most of all the domain of young men.

Riitta Lavikka (2002) and Päivi Korvajärvi (2002) have compared, from a gender perspective, work positions and work requirements in the manufacturing and service sectors, partly with cases described in this book. Their analyses are in line with earlier observations of the persistence of the gendered social order and divisions in Finnish working life (Rantalaiho and Heiskanen 1997). The differentiation of men's and women's jobs in such a way that women are in lower positions and get lower wages has not changed much, even though many other things have changed. The spread of information and communication technologies into the workplace and the increase of knowledge-intensiveness of work have not changed this basic fact, even with other changes in work and working conditions.

Lavikka concludes that traditional organizational culture seems to hinder major changes in gender-based task distributions and occupational boundaries, even though the meaning of these boundaries as an impediment to flexibility should be acknowledged. The thought and action models and traditions that make up organizational culture produce different positions and solutions for men and women. In spite of changes in technology, work organization and work content in the fields and case companies studied, technology was continually understood as a male area, whereas routines, support, service and human relations were seen as a female area. These observations on gendered divisions in working life in general, and relations to technology in particular, challenge universalizing talk about information society and show the need to include in the analysis the social orders and relations which shape actor positions and agency.

The case studies in this book have posed questions about space, technology, knowledge, expertise and work organization from an agency perspective, tracing lived experiences and meanings given to the situations in the new work settings by people in different actor positions.

Agency

Finland can be profiled as an information society not only through its highly developed technological infrastructure but also in terms of people's attitudes and everyday life. Attitudes towards technological development are very positive among Finnish people. The strong reliance on technology is visible alongside discussions on information society in daily technological practices. In this context, information systems (IS) specialists, who develop technology and mediate it to others, become core actors not only as shapers of technology but also as shapers of the information society.

The analysis in Tarja Tiainen's chapter in this volume on IS specialists' views on expertise, human, people and users has a substantial relevance from the point of view of human agency and action spaces created with and through technology. The study showed that quite often the IS specialists had very abstract views of people. Technology seemed changeable, whereas people were constructed as passive objects. The common usage of the term "users" in the IS field has a flavour of the same abstractness. The division between specialists and users tends to place specialists in the forefront and define users as "others" whose expertise is unimportant.

Marja Vehviläinen illuminates from another angle and with different terminology the place of technology in people's lives, showing ways in which the "others" could be at the forefront. In her case study we meet the social orders and relations within which technology is embedded, and citizens and their living worlds rather than simply users. Finland's governmental commitment to develop information society includes the formal aim of equal access to information technology as a universal right for citizens. From the agency perspective, equal access is not, however, enough. We need to know more about how and why people use or do not use information technology, how they integrate it into their activities, and what kinds of meanings they give it (cf. Wyatt 2001).

The study shows two different approaches to supporting access to information technology, approaches which started from different assumptions about the road to the information society and ended up in different methods. One approach aimed to secure, in line with national information society strategy documents, basic user skills in computing with standard examinations. The other approach took as its starting point the local information society, which, in contrast to the liberal theory behind the national strategy, emphasizes differences between different groups of citizens in their needs and interests as well as in their capabilities to use technology. This approach respected participants' own situations, their everyday lives, work and leisure, and strove to develop participants' understanding of technology in relation to their everyday activities. It is reasonable to assume that compared with the former approach, the latter is more likely to lead to an active shaping of technology instead of a mere adaptation to the demands of technology.

Tarja Tiainen's, Marja Vehviläinen's and Pernilla Gripenberg's studies all raise the question of expertise. In the IS specialists' case the question of expertise related to one professional group's special knowledge and power to use that knowledge, on the one hand in situations where other professional groups with different kinds of knowledge are involved, and on the other hand in society at large. In the local agency case, the question of expertise relates to citizens' rights and possibilities to acquire skills, which used to be the monopoly of computing professionals. With the former approach we are concerned with closed-context expertise, in the latter with open-context expertise (Giegel 1993). Concerns about specialized professional groups' ways of constructing expertise and expertise practices continue to be relevant issues in scientific discussion, but aside from such concerns, a discussion on expertise which is not the property of certain groups and institutions and which questions the division between experts and laypersons has gained increasing ground. In the information society context, this latter type of discussion helps to capture the expanding knowledge needs that cannot be met through existing institutions and expertise practices, and provides conceptual tools for taking a critical stance towards expertise.

Identity positions

The question of expertise provides a background to the discussion of knowledge work. In contrast to the rather scholarly talk about expertise, the knowledge work debate, firmly associated with the political information/knowledge society context, distinguishes social divisions more directly by defining the identity positions of different groups. Definitions of knowledge workers tell us who belongs to the core of the information society and who does not.

An essential part of the idea of the information society has been expected changes in occupational structures and work content. One attempt to move from visionary talk to a more practical level in describing the core trends in the development has been the construction of a classification system of information occupations (Porat 1998). Another attempt is the debate around the nature of knowledge work (e.g. Blackler 1995; Cortada 1998; Davenport and Prusak 1998; Alvesson 2001; Blom *et al.* 2001). Compared with the definition of information occupations (knowledge producers, distributors, processors), the definition of knowledge work ·is more concise, emphasizing expert-type work tasks that require creativity and innovativeness. Still, both definitions suffer from ambiguities in the concepts of information and knowledge as classification criteria. The cases in this book showed that innovation pressures and the need for knowledge-seeking and generation are quite frequent requirements in present-day organizations and have an influence on all personnel groups throughout the organization, not only specified knowledge workers.

While the basis for identity positions which the rhetorical divide between knowledge work/non-knowledge work draws is shaky, the increase of knowledge-intensiveness is itself of importance for the construction of identities. This question has been dealt with in relation to work, which typically would fall into the knowledge work category but is not limited to that only.

In researching identity construction in knowledge-intensive companies, Alvesson (2001) argues that knowledge itself is not the distinctive feature of such work. According to him, it is frequently impossible to separate knowledge and "pure" intellectual skills (symbolic analytical work) from flexibility, organizing capacity, a high level of motivation, social skills, less esoteric technical skills, the ability to follow company methods, and standardized ways of working (2001: 867). Quite often, uncertainty, complexity, instability and uniqueness fit the nature of such work better than the rational model of knowledge, or, as Alvesson puts it, where knowledge intensity is central, so is ambiguity. The construction of identities is, according to him, contingent upon these ambiguities, bringing forth issues of image, rhetoric and orchestration of social relations and processes.

Sirpa Kolehmainen's chapter (Chapter 5) also focused on knowledge-intensive companies, especially in the IT field. She deals with issues of control and commitment with the question: what are or would be organizational spaces that support both autonomous and shared knowledge creation and learning and strengthen the organizational commitment of employees? Her study and some other studies have shown that work commitment is inevitably intertwined with patterns of control in knowledge-intensive organizations.

Knowledge-intensive firms are specialized in solving problems for their client organizations, and the experts working there by necessity require the autonomy and authority to decide on appropriate procedures in their work. In such conditions, indirect forms of control that support the development of organizational commitment are emphasized, compared with organizations with traditional hierarchical structures. The IT firms studied worked on the basis of shared expertise, which relied both on peer control of the project group and on the self-control of an individual expert. The high expectations of self-control make different aspects of self-regulation important, and, contingent on these, different aspects of commitment: not only organizational commitment, but also work ethic, job involvement and career commitment.

Both ambiguity and the demands for simultaneous peer and self-control characterize work even more broadly than just in special knowledge work categories. We have talked of knowledge intensity in this book in conditions in which organizations face the need to innovate, solve problems and learn new things at an increasing pace. Such conditions may affect traditional manufacturing companies as well as new production fields and new types of organizations. The cases in this book showed a high degree of involvement in work in different occupational groups and a high degree of willingness to meet the requirements of continuous learning and development. High levels of involvement and commitment were reflected also in different occupational groups' willingness to be flexible concerning working hours, even up to the point where it is no longer reasonable from the point of view of their personal well-being.

Time, place and way of life

Behind the individual reactions to working hours are more general changes in time structures in developed societies. The industrial working time regime is changing from the disciplined system towards higher variation in working times and diffuse borders between working time and one's own time (Sennett 1998). Julkunen and Nätti (2000) assert that, through socialization processes, flexibility and adaptation to changing time needs of workplaces are becoming self-evident in the same way as adaptation to the disciplined time regimes of the industrial era. Global information networks of the new economy have tied countries in different parts of the globe in the same rhythm of functioning and certain characteristics of knowledge-intensive work have made it less dependent on time and place. Time structures intrude into self-processes. Casey (1995) and Hochschild (1997) have claimed that the centrality of work in one's life and marginalization of private life in the face of the requirements of work seem to be an essential orientation in knowledge and expert work. Julkunen and Nätti (2000: 202) have pointed out characteristics in knowledge-intensive work that tend to lead to a process in which work begins to colonize one's whole life. According to them, knowledge work is usually personal in the sense that no other person continues where the expert stops. It is also limitless; it can be done better, more and longer. It follows in thoughts, as data or text from the workplace to home, or it can be transferred electronically. The workplace can be at home or in the train, where portable PCs and mobile telephones can be taken along. Work is itself challenging and interesting, and also gives other rewards than just economic success. Such characteristics easily lead to behaviour patterns in which freedom from time and place constraints are used in favour of work.

In congruence with other studies and writings on changing time structures, long working hours and flexibility in the weekly rhythm of work characterized most clearly planning, design and expert work in our cases, but were not unknown to other occupational groups, either. In Finland working hours are institutionally relatively flexible. The old, relatively strictly regulated system of working time and the new flexible paradigm live side by side (Antila and Ylöstalo 1999). Workers in the new knowledge work areas have been able to create their own norms and practices concerning working hours.

In Kolehmainen's study, workers in knowledge-intensive business services made, from some perspective, individual choices on working hours, but they also made choices with an internalized awareness of time pressures resulting from time schedules of projects. In spite of the relative freedom of time and place constraints, the knowledge-intensive business services (KIBS) workers preferred regular working hours per week and working in the firm instead of at home or in the client's organization. These preferences can be seen in light of trying to maintain boundaries between work and private life in conditions where work tends to become presented as a way of life. A stronger version of this struggle can be seen in Chapter 9, on telework, where the lack of institutionally set time and place constraints is even clearer. When work becomes a state of mind in the absence of structure-providing external forms, as one of Riikka Kivimäki's interviewees described his orientation to work, time management becomes a key factor in the control of one's life.

Social requirements

The cases in the book represent very different kinds of jobs with different requirements. However, one common feature could be observed, namely the great significance of social requirements. The need for fluent co-operation and social skills was emphasized everywhere, even though the specific sources of these requirements were different.

Pernilla Gripenberg's chapter on the virtual office and virtualizing processes (Chapter 6) highlights how micro-level social interactions and social requirements were in many ways at the core of the change processes. These social processes in turn changed the relation of the social and the technological, and these changes were institutionalized in the re-forming of the organizations and their work process. The KIBS experts had to get along both with their peers and with the people in the client organization. The idea of shared expertise behind the organization of work tasks made the co-operation necessary, since in big assignments the part defined for one expert had an influence on others' work and on keeping to the overall schedule. The close ties between the KIBS firm and the client organization were felt also in the work of the individual worker, from the definition phase at the beginning of the project to the evaluation of the outcome. In the manufacturing companies, the co-operation requirements resulted from the team-based organization of work tasks, requirements of interaction across functional and hierarchical divisions within the company, placement of the company's units in geographically distant places and an increased requirement of sensitivity to the customers' needs, wishes and expectations, combined with the strict delivery times of the products and the ever-present time pressure. In the call centre, the social interaction was literally the core of the business, through which the company earned its profit and constructed its outward image.

Richard Sennett (1998) has argued that the modern work ethic focuses on team work. As he says (p. 99), it celebrates sensitivity to others; it requires such "soft skills" as being a good listener and being co-operative; and most of all, team work emphasizes team adaptability to circumstances. On the other hand, teamwork leads, according to him, to the absence of authority (pp. 109, 115). The emphasis on the soft skills of communication and mediation leaves the person in power in the role of enabler and not in the role of an authority who can say, "This is the right way."

This question of authority raised by Sennett is one of the confusing aspects of the increased requirements for co-operation and social skills, but there are others, too. The simultaneous emphasis on specialization in expert areas, on the one hand, and the requirement for multiskills, on the other hand, is one source of confusion. As a worker in one of the Finnish case companies studied, a metal firm, ironically remarked on the changing requirements of the job:

> Formerly, the high occupational skills were based on education, you were a welder, machinist, etc. Nowadays, one can say, one has to have all these [skills] and in addition to that, the skills of a technician, an engineer, and human relations skills. And further, one could almost say, the skills of a physical trainer, in order to maintain one's physical condition. So … this skill requirement, it has no upper limit.

To meet the requirements of profound, specialized competence, multiskillness and competence integration serving the entire organization, solutions have been developed in the shape of teams, project groups and co-operation between workers of different competences. Co-operation is felt and accepted as a necessary part of the work among different occupational groups, but it is in many cases co-operation with internal confusions and contradictions. An illustrative example of this is the co-operation between shop-floor workers and product designers. The metal firm had moved the production of the prototypes from a separate department to the shop floor amid the normal production. Workers, designers and managers found this to be a welcome change. Developing each prototype created a situation of intensive co-operation between the production and the planning. From a practical point of view, it was a very reasonable arrangement and avoided many quarrels, which earlier had been quite common when drawings and specifications did not match the practical possibilities to proceed according to the design. However, this kind of working requiring crossing customary organizational and cultural barriers, and even learning a new "language" was a gradual process rather than a once-and-for-all change of practices.

Spaces

Technology has provided new means for communication and new possibilities for organizing work activities and dividing work tasks. We could also say that technology has been a part of the arrangement of virtual working. The cases described different kinds of technologically mediated spaces. The IT competences which the local information society programme ensured opened up completely new channels of communication and a new source of information for the participants. The call centre applied the possibilities of the newest technology to organize communication. With the teleworkers, e-mail and mobile phones essentially created the link between the place where they worked and their host organization or clients. The experiment with project work between the research and development projects created a virtual organization for communication and co-operation.

Jackson (1999) has identified five main images and perspectives in discussions on virtual working. One line of discussion refers to the novel work configurations resulting from digitalizing or formatting work, through which representations of the world are encoded in computer software, allowing people to interact in a virtual world. Second, the discussion has raised the question of organizational flexibility, the need of the organization to evolve, redefine and reinvent itself continually for practical business purposes, including the aim to overcome the constraints related to time and space. Third, virtual working or the virtual organization is defined by the absence of a human component (colleagues, customers) and is thus discussed as a counterpoint to images embodied in offices and factories. Fourth, boundary erosion, either within or between organizations, and the continually changing interfaces among a company, suppliers and customers have been at the forefront in the discussion and definitions of virtual working and organizations. Fifth, perspectives and problems of electronic commerce have been touched on in the context of virtual working. While looking at these lines of discussion, one can observe that virtual working can refer to work configurations and aspects of work in different kinds of settings, both in new types of organizations and in more traditional organizations, such as in manufacturing companies, for example.

Communities and identification

The meaning of community ties and increased co-operation and social requirements has been discussed in the literature from different perspectives and with contradictory conclusions. Casey (1995) and Ezzy (2001), using examples from manufacturing industries, have addressed this issue in terms of view of employees' subjectivity, and connecting their analyses with the contested debate on individualism and communitarianism. Both of them draw attention to how the structural conditions of teamwork organization and the cultural aspects of a co-operative climate are integrated. Their analyses suggest that workers invest themselves in the workplace "family" through internalization of responsibility to their team while being simultaneously aware of the fact that the corporate culture of trust and co-operation is something in which they cannot completely believe. These conditions prepare the soil for seeking personal self-fulfilment through a kind of instrumental individualism. Even though the empirical basis from which Casey and Ezzy draw their conclusions is limited, their arguments raise thought-provoking lines of analysis for studying commitment, identification and agency. These issues generally remain not dealt with, however, in recent writings on knowledge management that tackle problems of social setting and community.

Matters of virtual working have often been approached in a rather technical manner, and with an optimistic, even evangelist tone, as Jackson (1999: 7) remarks. The cases in this book bring to the fore the social processes and organizational dynamics of such configurations. The experiment by Riitta Kuusinen, aimed at creating a space for co-operation between research/development projects, showed that a lot more must be taken into consideration than simply the material basis of the space. Gathering and modifying knowledge in different projects is one of the characteristic features of the present-day information society. Different forms of co-operation are needed for processing information within and across organizations. The experiment aimed at shared knowledge-processing among a large number of projects in different parts of Finland. The constructed web-forum provided the material basis for the co-operation. In the event, it proved that the actual knowledge management strategy was not in line with the espoused expectation of co-operation and shared knowledge-processing. In comparison to the set objectives, the space stayed empty and the participants continued their customary individual ways of working instead of engaging in interactive knowledge-processing.

The call centre, telework, KIBS and virtual office examples bring forth another aspect of social processes in new types of work settings, namely the importance of community. In the call centre, the working space for customer work was virtual, the community of co-workers tied to the concrete workplace. The workers found the social community important both because of the chance to reflect on work-related problems and because of socializing activities. The same reasons came up also in the interviews of teleworkers and KIBS workers. The KIBS workers emphasized especially the need to have discussions with colleagues during the problem-solving process. The teleworkers made special arrangements in order to maintain linkages to the work community.

Cornfield *et al.* (2001) start their book *Working in Restructured Workplaces* with a reference to the classic book *The Organization Man*, in which William H. Whyte (1956) announced the demise of the individualistic entrepreneur and the entry of the "organization man", who sought to belong to the large corporation through a lifetime career and mutual commitment with his employer. The authors raise the question of whether the workplace continues to constitute a community for organization men and women. Many changes since the publication of Whyte's book have reduced the room for manoeuvre of organization men and women: employment relationships have changed, new network-type organizational arrangements have appeared. The workplace is no longer the self-evident natural unit of belonging. Still, the cases in this book show that the desire for a work-related community has not disappeared. Quite the contrary, in fact.

Ezzy argues that identification with the company and its teams takes the place of occupational and professional groups. The development trends concerning work practices and their structural conditions, which our cases also exemplify, suggest that this might indeed be happening. Teams, project groups and activities crossing functional and hierarchical boundaries mix the composition of co-operation partners and the day-to-day contacts and may reduce links to one's own occupation or profession.

A notable aspect in the case examples was the identification with the images of the company concerning its prospects. These images are also deliberately used by management in the recruitment policies. The interviews with the directors of the KIBS firms are revealing in this sense:

> When interviewing new candidates for a job, I used to ask them what was the most important thing for them in work. Most candidates usually answered that it was intrinsically rewarding work, and second, the chance that it offered to learn new things, all the time.

Another interviewee stated:

> These workers are really technology buffs. All the time there is something new.

The KIBS workers' positive orientation to technology is not surprising. In the case of the call centre, where the positive orientation is to be less self-evidently expected, technology as an object of identification seems to need a broader contextual interpretation.

Attitudes towards technological development are generally very positive among Finnish people. Castells and Himanen (2001) have pointed out that Finnish people's identity is future-oriented and that the collective achievements of the information society, especially the breakthroughs in technological development, strengthen this identity. Korvajärvi has interpreted the call centre workers' orientation to technology specifically within this kind of contextual framework. The positive attitudes among the call centre workers could be understood through the well-being guarantee that the technology seemed to promise. In some respects, technology was in the background in their experiences, as a self-evident part of the customer service. On the other hand, technology was admired and in that sense took centre-stage in their experiences. Technology was to them something that not only guaranteed the existence of the firm and the jobs but also represented the firm with its promises.

Just as identification with companies' prospective images deserves attention, so does the preparedness of the workers to invest themselves in the process of achieving the progressive results. All the occupational groups valued development activities in their workplaces, even though sometimes with a hint of sarcasm when development "isms" one after the other marched in and out. They also valued the possibility to develop their competences and showed willingness to learn new things. It is possible that the interpretation of instrumental individualism, made, for example, by Casey and Ezzy, applies also to these orientations. This is not, however, an exhaustive explanation of being in the workplaces. There is, for example, still something authentic in the desire for belonging to the workplace community.

Beyond the information society

Transformations in work and employment structures, work processes and the quality of working life are typically discussed in the context of broader social and economic change, which nowadays is often conceptualized with the notion of information society. There tends to be a universalistic tone in these discussions, as if the information society would be something similar everywhere. This book seeks to remind readers of the differences and choices in the path towards the information society. It is true that there are some development trends, such as increasing globalization, informatization, expanding use of information and communication technologies in different spheres of society, and widening application of the network principle in the organization of activities that legitimize the term "information society" in different societies. The cases in this book come from a specific kind of information society, which has integrated aspects of the welfare state and the information society. The development of the model has taken place as a gradual process, as a consequence of a deliberate policy in which social and technological concerns are consciously integrated. Still, it would be an exaggeration to claim that the social and technological matters are settled in the Finnish model of the information society. The Finnish future researchers Markku Wilenius and Matti Kamppinen (2000: 14) proclaim in their article "Living beyond the information society" that instead of striving for an information society in which the development of information technology is the primary social goal, we should aim to make the world a networked community of people capable of positive interaction. This wish is also valid from the point of view of working life. The development of information society provides new possibilities for work configurations and employment relations. An agency perspective starting from human aspirations helps to ensure that the much sought-after forms of flexibility would be sustainable in terms of people's well-being.

Postscript

Information societies are still societies

Jeff Hearn

Information society or information societies are still *societies*. Indeed, as is so often the case with social science concepts, it is necessary to now speak of "the information society" *in the plural*. This means speaking of specific kinds of information society or rather information societies. Just as there are different kinds of capitalism and patriarchy, so there are different kinds of information society. Information societies can differ both in relation to configurations of capital, the market, state, welfare and civil society, as well as social divisions, such as gender, ethnicity and racialization. Social divisions pervade any universalizing tendencies in society. There are thus quite different possible trajectories of information society, rather than a single convergent model.

And clearly, being societies, information societies retain all the features of society: power relations, authority, inequalities, hierarchies, inclusion and exclusion, social networks, and so on. In information societies, people still sleep, get up, eat, work for no pay or for pay, cook, clean, have friends and enemies, have sexual relations, use violence, stay well, get ill, grow older, die, and do all the things that people have done before. While all these activities may change to some extent within information societies, people also do a few more things, too, in living and working with information and information technology. Information society, like virtuality, adds to the sum of human activities. Social life and sociality continue.

Information

In the preface to this book, I suggested that in a critical cultural perspective on information in organizations and workplaces in information society, *information – what counts as information – is context-specific socially and societally*. While there have been any number of macro-level (and often somewhat speculative) analyses of information societies, we are only at the beginning of understanding what is happening "on the ground".

There is still a lack of micro-level studies of what people are doing in information society, day to day. It is still far from clear what is happening at the level of social practice, and what the long-term implications of these changes towards informatization may be. Thus this book has addressed these questions through *the everyday life of information society, by detailed, ethnographic and qualitative research.* From this basis, the need for analysis of the grounded and sometimes contradictory processes of different social changes throughout different information societies now seems all the stronger.

And information societies do clearly bring changes: some obvious, for example, in the use of technology and creation of very large databases, some much less so. One aspect of this differencing and this contextualizing is the many different kinds of information that there can be. There is a danger in talking of information within the context of information society, as if it is one thing. Information can take a huge number of physical and social forms. This focus also highlights shifts from "knowledge" (a certain valuation) to "information" (a certain form of knowledge – information knowledge), and thence to "virtuality" (a certain form of knowledge and information – virtual information knowledge). There is no more reason why information should be one thing than, say, art or science or education might be. Accordingly, the social practices, social time(s) and social space(s) developing around information and its production and processing are similarly diverse. In the information society, the contextuality of information stays in a complex relation to its apparent decontextualization (Strathern 2002).

Information technology

In information society, *information is increasingly a computer-mediated production and reproduction.* However, studying this not only concerns what happens in computer use or in computer-mediated communication, but developing greater understanding of different relations of such uses throughout society. This has involved addressing the *everyday, cultural, local experiences of and agentic practices around information technology.*

Information society and the use of information and communication technologies (ICTs) are not exact equivalents. However, it is clear that a key aspect of change towards what might be called information society is the development of ICTs. They involve the use of multiple complex technologies and have several characteristic features. These include: time/space compression, instantaneousness, asynchronicity, reproducibility of image production, the creation of virtual bodies, surveillability, the blurring of the real and the representational. ICTs can be understood as both providing "free space" in local networks unfettered by centralized control or moral codes, and the most surveilled social arena yet, including in surveillance in particular work organizations.

Importantly, these technologies are not to be understood as just texts but exist within and indeed create material social relations. To use the terminology adopted in the original formulation of the research project which spawned most of these chapters and case studies, these are matters of shifting, mobile social boundaries. Such mobile boundaries appear to take many forms, between: humans and technology, home and work, work and leisure, management and labour, public and private (in both senses), "social reality" and social representation, virtuality and embodiment, and so on.

Workplaces

This brings us on to the question of the workplace. Workplaces can be seen as the largest producers and processors of information, even with the diffusion of informational modes throughout many and various social spheres other than the workplace. Yet *information in information society can be understood as increasingly produced from beyond the organization and organizational workplaces.*

So what is the changing place of the workplace in information societies? How does all this relate to the changing shape of work and the workplace(s)? Workplaces and organizations are characterized by information that transcends the organization and workplace. Information society involves an increasing focus within society on organizations and workplaces that are characterized by their own transcendence. Thus the task here has been to focus on *everyday, cultural experiences of information technology in workplaces and organizations.* Organizations and workplaces may appear to be increasingly abstracted, disembodied and defined by information given meaning from beyond themselves, but the cultural experiences of work remain embodied.

There may appear to be profound changes in process in the very constructions and interrelations of social divisions – age, class, ethnicity, gender, and so on – and their interconnections with work and workplaces. It seems increasingly difficult to discuss any of these in isolation from each other. In one sense, one could argue that this has always been so historically, it is just that it has not been noticed so much until rather recently. Social changes towards information societies also seem to contribute to the increasing development of and elaboration in intersectionalities between social divisions. It might also be that the very formation of "people" as persons, bodies, individuals may be in the process of profound historical change, in part through the impact of information societies. Rather than people being formed primarily as fixed embodied *members* of given collectivities, defined by single social divisions, people may increasingly appear to exist and be formed in the social relations, spaces and practices *between* multiple power differentials. Persons and bodies are no longer so easily equated. The relation of social class and work positioning is no longer so clear-cut, even though most societies remain clearly capitalist and patriarchal. Information society exists in workplaces, as fragmented hegemonies (see Poster 2002) and mini-globalities.

Society

A final paradoxical theme is that while information is increasingly from beyond a given society, the focus in studying these matters has been on a particular society, Finland, the society that can most strongly claim to be an information society. Despite the universalizing tendencies and presumptions of movements towards information society/ies, the studies presented are all based in the local, cultural milieux of Finland.

Informationization is intimately linked to globalization, global capitalism and desocietalization, which is not only a political process of fragmentation of nation-states, but a cultural process, with the transcending of boundaries by information and cultural artefacts. Though information societies are societies, this does not mean that the notion of "society" has to be limited to the nation-state, as has been so for much of the study of different societies. Information societies are not linked so directly to the specifics of geographical place as previously, even though they still exist in places and social spaces. While desocietalization emphasizes the transcendence of the nation-state, the increasing importance of signs, symbols and transnational cultures, there is also, paradoxically, a need to emphasize the ways that both nation-states and organized labour remain important within political economy (Gibson-Graham 1996; Edwards and Elger 1999; Waddington 1999). Even with contemporary economic, cultural and other forms of globalizing imperialism, especially but not only of the US, specific societies, cultures and nations remain important.

Look no further than the place where you are to examine these apparently placeless social changes, practices and experiences at work. For my own part, living most of my time in Finland, away from my country of origin, the UK, has led me to become more interested in questions of context and culture. I may say to myself: "Things are different here," but what is meant by this is sometimes difficult to elucidate. To take one specific linguistic example: in Finnish the words for computer, information technology and information society are respectively *tietokone* ("knowledge machine"), *tietotekniikka* ("knowledge technology") and *tietoyhteiskunta* ("knowledge society"). If we talk of information society, it certainly takes a different form in the two countries I know best.

This book has sought to contribute to critical cultural studies on information societies. It has done this by focusing on people's everyday relations to information technology in and around organizational workplaces. This focus on workplaces is important, even though, paradoxically, information can be understood as increasingly produced from beyond the organization and organizational workplaces. In that way organizations and workplaces can be said to be diffused and transcended in information societies. And even though information in workplaces is also increasingly from beyond a given society, these studies have been in Finland, as paradoxically both *a* specific national location and arguably *the* information society.

References

Acker, J. (1992) "Gendering organizational theory", in A.J. Mills and P. Tancred (eds) *Gendering Organizational Analysis*, Newbury Park: Sage.

Adami, L.M. (1999) "Autonomy, control and the virtual worker", in P. Jackson (ed.) *Virtual Working, Social and Organisational Dynamics*, London: Routledge.

Agarwal, R., Sambamurthy, V. and Stair, R.M. (1997): "Cognitive absorption and the adoption of new information technologies", *Academy of Management Conference Proceedings*, Boston.

Agres, C., Edberg, D. and Igbaria, M. (1998) "Transformation to virtual societies: forces and issues", *Information Society*, 14, 2: 71–82.

Aho, S., Kivimäki, R. and Koski, P. (1995) *Uusi kestävä yrittäjyys*, Työministeriön julkaisuja, Helsinki: Työministeriö.

Alasoini, T. (1999) "Organizational innovations as a source of competitive advantage – new challenges for Finnish companies and the National Workplace Development Infrastructure", in G. Schienstock and O. Kuusi (1999) *Transformation Towards a Learning Economy: the challenge for the Finnish Innovation System*, Sitra 213, Helsinki: Sitra.

Althusser, L. (1971) "Ideology and ideological state apparatuses (Notes towards investigation)", in L. Althusser *Lenin and Philosophy and Other Essays*, London: New Left Books.

Alvesson, M. (1993) "Organizations as rhetoric: knowledge-intensive firms and the struggle with ambiguity", *Journal of Management Studies*, 30, 6: 997–1015.

—— (2000) "Social identity and the problem of loyalty in knowledge-intensive companies", *Journal of Management Studies*, 37, 8: 1101–23.

—— (2001) "Knowledge work: ambiguity, image, identity", *Human Relations*, 54, 7: 863–86.

Alvesson, M. and Karreman, D. (2000) "Varieties of discourse: on the study of organizations through discourse analysis", *Human Relations*, 53, 9: 1125–49.

Anderson, J.R. (1996) *The Architecture of Cognition*, Mahwah, New Jersey: Lawrence Erlbaum Associates.

Anderson, R.E., Johnson, D.G., Gotterbarn, D. and Perrolle, J. (1993) "Using the new ACM Code of Ethics in decision making", *Communications of the ACM*, 36, 2: 98–105.

Antila, J. and Ylöstalo, P. (1999) *Enterprises as Employers in Finland: Flexible Enterprise Project*, Työpoliittisia tutkimuksia 205, Helsinki: Ministry of Labour.

Argyris, C. (1993) "Education for leading-learning", *Organizational Dynamics*, 21, 3: 5–17.

Argyris, C. and Schön, D.(1975) *Theory in Practice: Increasing Professional Effectiveness*, San Francisco: Jossey-Bass.

—— (1978) *Organizational Learning: a Theory of Action Perspective*, Reading, Mass: Addison Wesley.

Aro, J. (2000) "Tietoteknologinen kehitys ja yhteiskunnallinen muutos", in M. Vuorensyrjä and R. Savolainen (eds) *Tieto ja tietoyhteiskunta*, Helsinki: Gaudeamus.

— (2001) "Narratives and rhetoric of the information society in administrative programs and in popular discourse", in E. Karvonen (ed.) *Informational Societies, Understanding the Third Industrial Revolution*, Tampere: Tampere University Press.

Atkins, M. and Dawson, P. (2001) "The virtual organization: emerging forms of ICT-based work arrangements", *Journal of General Management*, 26, 3: 41–52.

Audi, R. (ed.) (1995) *The Cambridge Dictionary of Philosophy*, New York: Cambridge University Press.

Avgerou, C. (2000) "Information systems: what sort of science is it?" *Omega*, 28, 5: 567–79.

Bagnara, S. (2000) *Towards Telework in Call Centres*, Euro-Telework, EC DG Employment and Social Affairs in the European Social Fund. Available HTTP: http://www.tele-work-mirti.org/bagnara.htm

Bandura, A. (1995) "Exercise of personal and collective efficacy in changing societies", in A. Bandura (ed.) *Self-efficacy in Changing Societies*, Cambridge: Cambridge University Press.

Barret, R. (2001) "Labouring under an illusion? The labour process of software development in the Australian information industry", *New Technology, Work and Employment*, 16, 1: 18–34.

Bauman, Z. (1995) *Life in Fragments: Essays in Postmodern Morality*, Oxford: Blackwell.

Beck, U. and Beck-Gernsheim, E. (2002) *Individualization, Institutionalized Individualism and its Social and Political Consequences*, London: Sage.

Beirne, M., Ramsay, H. and Panteli, A. (1998) "Developments in computing work: control and contradiction in the software labour process", in P. Thompson and C. Warhurst (eds) *Workplaces of the Future*, Wiltshire: Macmillan.

Bell, D. (1973) *The Coming of Post-industrial Society: a Venture in Social Forecasting*, New York: Basic Books.

Bereiter, C. (2000) *Education and Mind in the Knowledge Age*. Online. Available HTTP: http://csile.oise.utoronto.ca/edmind/main.html

Bereiter, C. and Scardamalia, M. (1993) *Surpassing Ourselves: an Inquiry into the Nature and Implications of Expertise*, Peru, Illinois: Open Court.

Bijker, W.E. and Law, J. (eds) (1992) *Shaping Technology/Building Society*, Cambridge Mass: MIT Press.

Bijker, W.E., Hughes, T.P. and Pinch, T. (eds) (1987) *The Social Construction of Technological Systems*, Cambridge, Mass.: MIT Press.

Bjerknes, G. and Bratteteig, T. (1995) "User participation and democracy: a discussion of Scandinavian research on system development", *Scandinavian Journal of Information Systems*, 7, 1: 73–98.

Björkegren, C. and Rapp, B. (1999) "Learning and knowledge management: a theoretical framework for learning in flexible organisations", in P. Jackson (ed.) *Virtual Working*, London: Routledge.

Bjørn-Andersen, N. (1988) "Are 'human factors' human?" *The Computer Journal*, 31, 5: 386–90.

Blackler, F. (1995) "Knowledge, knowledge work and organizations: an overview and interpretation", *Organization Studies*, 16,6: 1021–46.

Blom, R., Melin, H. and Pyöriä, P. (eds) (2001) *Tietotyö ja työelämän muutos: Palkkatyön arki tietoyhteiskunnassa*, Gaudeamus Kirja / Oy Yliopistokustannus University Press Finland, Tammer-Paino: Tampere, Finland.

Brancheau, J.C. and Brown, C.V. (1993) "The management of end-user computing: status and directions", *ACM Computing Surveys*, 25, 4: 437–82.

Brooking, A. (1999) *Corporate Memory, Strategies for Knowledge Management*, London: International Thomson Business Press.

Brown, J.S. and Duguid, P. (2000) "Balancing act: how to capture knowledge without killing it", *Harvard Business Review*,78, 3: 73–80.

Burkhardt, M.E. (1994): "Social interaction effects following a technological change: a longitudinal investigation", *Academy of Management Journal*, 37, 4: 869–98.

Burkhardt, M.E. and Brass, D.E. (1990): "Changing patterns or patterns of change: the effects of a change in technology on social network structure and power", *Administrative Science Quarterly*, 35: 104–27.

By Joint Work Party to the Information Society: The Information Society Strategy and Action of North Karelia 1999–2006 (1999), Joensuu: The Regional Council of North Karelia.

Byrd, T. and Marshal, T. (1997): "Relating information technology investment to organizational performance: a causal model analysis", *International Journal of Management Science*, 25, 1: 43–56.

Callaghan, G. and Thompson, P. (2001) "Edwards revisited: technical control and call centres", *Economic and Industrial Democracy*, 22, 1: 13–37.

Callon, M. (1986) "Some elements of a sociology of translation: domestication of the scallops and the fisherman of St Brieuc Bay", in J. Law (ed.) *Power, Action and Belief: a New Sociology of Knowledge?* London: Routledge and Kegan Paul.

Callon, M. and Latour, B. (1981) "Unscrewing the big Leviathan: how actors macrostructure reality and how sociologists help them do so", in K. Knorr and A. Cicourel (eds) *Advances in Social Theory and Methodology: Toward an Integration of Micro- and Macrosociologies*, Boston, London and Henley: Routledge and Kegan Paul.

— (1992) "Don't throw the baby out with the bath school, a reply to Collins and Yearley", in A. Pickering (ed.) *Science as Practice and Culture*, Chicago: Chicago University Press.

Cappelli, P. (1999) *The New Deal at Work*, Boston Mass.: Harvard Business School Press.

Carnoy, M. (2000) *Sustaining the New Economy, Work, Family, and Community in the Information Age*, Cambridge Mass.: Harvard Unversity Press.

Casey, C. (1995) *Work, Self and Society after Industrialism*, London and New York: Routledge.

— (2002) *Critical Analysis of Organizations: Theory, Practice, Revitalization*, London: Sage.

Castells, M. (1996) *The Rise of the Network Society*, Oxford: Blackwell.

— (1997) *The Power of Identity*, Volume II of *The Information Age, Economy, Society, Culture*, Oxford: Blackwell.

— (1998) *End of Millennium*, Oxford: Blackwell.

— (2000) "Information technology and global capitalism", in W. Hutton and A. Giddens (eds) *On the Edge: Living with Global Capitalism*, London: Jonathan Cape.

Castells, M. and Himanen, P. (2001) *The Finnish Model of the Information Society*, Sitra Reports Series 17, Vantaa: Sitra.

Clement, A. (1994) "Computing at work: empowering action by low-level users", *Communications of the ACM*, 37, 1: 52–65.

Collins, H.M. and Yearley, S. (1992) "Epistemological chicken", in A. Pickering (ed.) *Science as Practice and Culture*, Chicago: Chicago University Press.

Compaine, B.M. (2001) "Information gaps: myth or reality", in B.M. Compaine (ed.) *The Digital Divide. Facing a Crisis or Creating a Myth?* Cambridge, Mass.: MIT Press, 105–18.

Cooper, R. and Fox, S. (1990) "The texture of organizing", *Journal of Management Studies*, 27, 6: 575–82.

Cornfield, D.B., Campbell, K.E. and McGammon, H.J. (2001) *Working in Restructured Workplaces: Challenges and New Directions for the Sociology of Work*, London, New Delhi, Thousand Oaks: Sage.

Cortada, J.W. (ed.) (1998) *The Rise of the Knowledge Worker*, Boston: Butterworth-Heinemann.

Cronberg, T. (1999) "Pohjois-Karjala tietoyhteiskuntaan: alueelliset. toimijaverkot ja syrjäytymättömyyden rakentaminen" (North Karelia towards the information society: regional actor networks, building non-displacement, in Finnish), in P. Eriksson and M. Vehviläinen (eds) *Tietoyhteiskunta seisakkeella: teknologia, strategiat ja paikalliset tulkinnat*, SoPhi, Jyväskylä, 215–30.

Crozier, M. (1964) *The Bureaucratic Phenomenon*, Chicago: University of Chicago Press.

Csordas, T. (2000) "Introduction: the body as representation and being-in-the-world", in T. Csordas (ed.) (2000) *Embodiment and Experience: the Existential Ground of Culture and Self*, Cambridge: Cambridge University Press.

Dahlbom, B. and Mathiassen, L. (1997) "The future of our profession", *Communications of the ACM*, 40, 6: 80–9.

Davenport, T.H. and Prusak, L. (1998) *Working Knowledge: How Organizations Manage What They Learn*, Boston: Harvard Business School Press.

Davidson, A.L., Schofield, J. and Stock, J. (2001) "Professional cultures and collaborative efforts: a case study of technologists and educators working for change", *The Information Society*, 17, 1: 21–32.

Deci, E.L. and Ryan, R.M. (1987) "The support of autonomy and the control behavior", *Journal of Personality and Social Psychology*, 53, 6: 1024–37.

Demaret, L., Quinn, P. and Grumiau, S. (1998) *Call Centres: the New Assembly Lines*. Online. Available HTTP: http://www.labournet.org.uk/1998/August/call.html

Denning, P.J. (2001) "Who are we?" *Communications of the ACM*, 44, 2: 15–19.

Depickere, A. (1999) "Managing virtual working", in P. Jackson (ed.) *Virtual Working: Social and Organisational Dynamics*, London: Routledge.

Di Martino, V. and Wirth, L. (1990) "Telework: a new way of working and living", *International Labour Review*, 129, 5: 529–54.

Drucker, P.F. (1969) *The Age of Discontinuity: Guidelines to Our Changing Society*, London: Heinemann.

Durkheim, E. (1964) *The Division of Labour*, New York: Free Press.

Edwards, P. and Elger, T. (eds) (1999) *The Global Economy, National States, and the Regulation of Labour*, London: Mansell.

Ehn, P. (1988) *Work-oriented Design of Computer Artifacts*, Stockholm: Arbetslivscentrum.

Eriksson, P. (1999) "… On aika tehdä lopullinen ratkaisu … Strateginen johtaminen ja osallistuminen tietoyhteiskunnan kaupungeissa", in P. Eriksson and M. Vehviläinen (eds) *Tietoyhteiskunta seisakkeella: Teknologia, strategiat ja paikalliset tulkinnat*, Jyväskylä: Sophi.

Eriksson, P. and Vehviläinen, M. (eds) (1999) *Tietoyhteiskunta seisakkeella, Teknologia, strategiat ja paikalliset tulkinnat*, Jyväskylä: Sophi.

Escobar, A. (1999) "Gender, place and networks: a political ecology of cyberculture", in W. Harcourt (ed.) *Women @ Internet: Creating New Cultures in Cyberspace*, London: Zed Books, 31–54.

Eteläpelto, A. (1994) "Work experience and the development of expertise", in W.J. Nijhof and J.N. Streumer (eds) *Flexibility and Training in Vocational Education*, Utrecht: Lemma.

Etzioni, A. (1961) *A Comparative Analysis of Complex Organizations: On Power, Involvement and Their Correlates*, Glencoe: Free Press.

EVA (1999) *EVA-raportti: Suomalaisten asenteet 1999*, Helsinki: EVA.

Ezzy, D. (2001) "A simulacrum of workplace community: individualism and engineered culture", *Sociology*, 35, 3: 631–50.

Fairclough, N. (1995) *Media Discourse*, London: Edward Arnold.

Fay, B. (1976) *Social Theory and Political Practice*, London: Allen & Unwin.

Fineman, S. (2000) "Emotional arenas revisited', in S. Fineman (ed.) *Emotion in Organizations*, London: Sage.

Foster, H. (ed.) (1985) *Postmodern Culture*, London: Pluto.

Frenkel, S., Korczynski, M., Donoghue, L. and Shire, K. (1995) "Re-constituting work: trends towards knowledge work and info-normative control', *Work, Employment and Society*, 9, 4: 773–96.

Frenkel, S.J., Korczynski, M., Shire, K.A. and Tam, M. (1999) *On The Front Line: Organization of Work in the Information Economy*, Ithaca and London: Cornell University Press.

Friedman, A.L. and Cornford, D.S. (1989) *Computer Systems Development: History, Organization and Implementation*, Chichester: John Wiley and Sons.

Gasser, L. (1986) "The integration of computing and routine work", *ACM Transactions on Office Information Systems*, 4, 3: 205–25.

Gherardi, S. (1995) *Gender, Symbolism and Organizational Cultures*, London: Sage.

Gibson-Graham, J.K. (1996) *The End of Capitalism (As We Knew It): A Feminist Critique of Political Economy*, Oxford: Blackwell.

Giddens, A. (1984) *The Constitution of Society: Outline of the Theory of Structuration*, Berkeley CA: University of California Press.

— (1989) "A reply to my critics", in D. Held and J.B. Thompson (eds) *Social Theory of Modern Societies: Anthony Giddens and His Critics*, Cambridge: Cambridge University Press.

— (1990) *The Consequences of Modernity*, Stanford: Stanford University Press.

— (1991) *Modernity and Self-Identity: Self and Society in the Late Modern Age*, Cambridge: Polity.

Giegel, H.-J. (1993) "Kontextneutralisierung und Kontextoffenheit als Strukturbedingungen der Gesellschaftlichen Riskokommunikation", in W. Bonss, R. Hohfeld and R. Kollek (eds) *Wissenschaft als Kontext – Kontexte der Wissenschaft*, Hamburg: Junius Verlag.

Goffman, E. (1963) *Encounters: Two Studies in the Sociology of Interaction*, University of California, Indianapolis, Indiana: Bobbs Merrill.

— (1974) *Frame Analysis: An Essay on the Organization of Experience*, Boston: Northeastern University Press.

Gorz, A. (1972) "Technical intelligence and the capitalist division of labour", *Telos*, 12: 27–42.

Greenbaum, J. and Kyng, M. (1991) "Introduction: situated design", in J. Greenbaum and M. Kyng (eds) *Design at Work: Cooperative Design of Computer Systems*, Hillsdale, NJ: Lawrence Erlbaum Associates.

Gregory, D. (1989) "Presences and absences: time–space relations and structuration theory", in D. Held and J.B. Thompson (eds) *Social Theory of Modern Societies: Anthony Giddens and His Critics*, Cambridge: Cambridge University Press.

Griffith, T.L. and Northcraft, G.B. (1993) "Promises, pitfalls, and paradox: cognitive elements in the implementation of new technology", *Journal of Managerial Issues*, 5, 4: 465–82.

Grint, K. (1998) *The Sociology of Work*, 2nd edition, Oxford: Polity.

Grint, K. and Woolgar, S. (1997) *The Machine at Work: Technology, Work and Organization*, Cambridge: Polity Press.

Gripenberg, P. (2002) "Living with IT: uses and interpretations of computers in the home and family context", *Proceedings of the Xth European Conference on Information Systems*, 6–8 June 2002, Gdansk, Poland, 1261–72.

Gripenberg, P. and Skogseid, I. with Botto, F., Silli, A. and Tuunainen, V.K. (2003) "Entering the European information society: 4 rural area development projects", *The Information Society*.

Grossberg, L. (1988) "It's a sin: politics, postmodernity and the popular", in L. Grossberg, T. Fry, A. Curthoys and P. Patton (eds) *It's a Sin: Essays on Postmodernism, Politics and Culture*, Sydney: Power Publications. Cited in L. McRae (2002) *Questions of Popular Culture*, Doctoral Thesis. Perth: Murdoch University.

Hacker, S. (1989) *Pleasure, Power and Technology*, Boston: Unwin Hyman.

— (1990) "The eye of the beholder: an essay on technology and eroticism", in S. Hacker, *Doing it the Hard Way*, edited by D.E. Smith and S.M. Turner, Boston: Unwin Hyman.

Hakkarainen, K., Lonka, K. and Lipponen, L. (2000) *Tutkiva oppiminen: älykkään toiminnan rajat ja niiden ylittäminen*, Porvoo: WSOY.

Hall, S. (1999) *Indentiteetti*, Tampere: Vastapaino.

Hamblin, H. (1995) "Employees' perspectives on one dimension of labour flexibility: working at distance", *Work, Employment and Society*, 9, 3: 437–98.

Håpnes, T. and Sørensen, K.H. (1995) "Competition and collaboration in male shaping of computing: a study of a Norwegian hacker culture", in K. Grint and R. Gill (eds) *The Gender–Technology Relation: Contemporary Theory and Research*, London: Taylor and Francis.

Haraway, D. (1991) *Simians, Cyborgs, and Women: The Reinventions of Nature*, London: Free Associations Books.

Hassard, J., Holliday, R. and Willmott, H. (eds) (2000) *Body and Organization*, London: Sage.

Haukness, J. (1998) *Services in Innovation: Innovation in Services*, S14S Final Report, Step Group. Online. Available HTTP: http://www.step.no

Hautamäki, A. (ed.) (1996) *Suomi teollisen ja tietoyhteiskunnan murroksessa, Tietoyhteiskunnan sosiaaliset ja yhteiskunnalliset vaikutukset*, Sitra 154, Helsinki: Sitra. Online. Available HTTP: http:// www.vn.fi/vn/vm/tyk/sitra154.html

Hearn, J. and Roberts, I. (1976) "Planning under difficulties: the move to decrementalism", in K. Jones (ed.) *The Yearbook of Social Policy in Britain 1975*, London: Routledge and Kegan Paul.

Heidegger, M. (2000) *Oleminen ja aika*, Tampere: Vastapaino.

Heikkilä, A. (1995) *Etätyön lailliset ja sopimukselliset perusteet*, Työpoliittinen tutkimus, Helsinki: Työministeriö.

Hendrickson, A.R. and Collins, M.R. (1996) "An assessment of structure and causation of IS usage", *DATA BASE*, 27, 2: 61–7.

Henwood, F., Plumeridge, S. and Stepulevage, L. (2000) "A tale of two cultures? Gender and inequality in computer education", in S. Wyatt, F. Henwood, N. Miller and P. Senker (eds) *Technology and In/equality: Questioning the Information Society*, London: Routledge.

Hirschheim, R. and Newman, M. (1988) "Information systems and user resistance: theory and practice", *The Computer Journal*, 31, 5: 398–408.

Hirschhorn, L. (1988) *The Workplace Within: The Psychodynamics of Organizational Life*, Cambridge, Mass.: MIT Press.

Hirschhorn, L. and Mokray, J. (1992) "Automation and competency, requirements in manufacturing: a case study", in P. Adler (ed.) *Technology and the Future of Work*, New York, Oxford: Oxford University Press.

Hirschman, A.O. (1970) *Exit, Voice and Loyalty: Responses to Decline in Firms*, Cambridge, Mass.: Harvard University Press.

Hochschild, A.R. (1983) *The Managed Heart: Commercialization of Human Feeling*, Berkeley, Los Angeles, London: University of California Press.

— (1997) *The Time Bind: When Work Becomes Home and Home Becomes Work*, New York: Metropolitan Books, Henry Holt.

Hollway, W. (1991) *Work Psychology and Organizational Behaviour: Managing the Individual at Work*, London: Sage.

Hollway, W. and Jefferson, T. (2000) *Doing Qualitative Research Differently: Free Associations, Narrative and the Interview Method*, London: Sage.

Huws, U. (1995) *Teleworking, Follow-up to the White Paper, Report to the European Commission's Employment Task Force* (Directorate General V) September, 1994, in *Social Europe, Follow-up to the White Paper, A-Teleworking, B-The informal sector*, Supplement 3/1995, Brussels/Luxemburg: European Commission, Directorate General for Employment, Industrial Relations and Social Affairs.

Igbaria, M. (1999) "The driving forces in the virtual society", *Communications of the ACM*, 42, 12: 64–70.

Ikonen, T. (2001) "Muistikuvia Marjalasta maailmalle" (Memories from Marjala out to the world), in J. Uotinen, S. Tuuva, M. Vehviläinen and S. Knuutila (eds) *Verkkojen kokijat paikallista tietoyhteiskuntaa tekemässä*, Suomen kansantietouden Tutkijain seura: Joensuu.

Isomäki, H. (1999) "Ontot tarinat: tietojärjestelmäammattilaisten ihmiskäsityksiä", in P. Eriksson and M. Vehviläinen (eds) *Tietoyhteiskunta seisakkeella: teknologia, strategiat ja paikalliset tulkinnat*, Jyväskylä: Sophi.

Ives, B., Hamilton, S. and Davis, G.B. (1980) "A framework for research in computer-based management information systems", *Management in Science*, 26, 9: 910–34.

Jackson, P.J. (1999) "Introduction: from new designs to new dynamics", in P.J. Jackson (ed.) *The Virtual Working: Social and Organisational Dynamics*, London and New York: Routledge.

James, P. (2001) "Abstracting modes of exchange: gifts, commodities and money", *Suomen Antropologi*, 2/2001: 4–22.

Jameson, F. (1991) *Postmodernism or, the Cultural Logic of Late Capitalism*, London: Verso.

Jones, S.G. (1995) "Understanding community in the information age", in S.G. Jones (ed.) *Cybersociety*, London, New Delhi, Thousand Oaks: Sage, 10–35.

Julkunen, R. (1987) *Työprosessi ja pitkät aallot: työn uusien organsaatiomuotojen synty ja yleistyminen*, Jyväskylä: Vastapaino.

— (2001) "Ammatti jälkiammatillisessa työelämässä", *Ammattikasvatuksen aikakausikirja*, 3, 2: 16–23.

Julkunen, R. and Nätti, J. (2000) "Uudet työkulttuurit, työaika, perhe ja sosiaalinen elämä", *Työ ja ihminen*, 14, 2: 198–205.

Karasti, H. (2001) *Increasing Sensitivity towards Everyday Work Practice in System Design*, doctoral dissertation, University of Oulu, Department of Information Processing Science, Oulu: Acta Universitatis Ouluensis A362.

Karvonen, E. (2000a) "Elämmekö tieto- vai informaatioyhteiskunnassa?" in M. Vuorensyrjä and R. Savolainen (eds) *Tieto ja Tietoyhteiskunta*, Helsinki: Gaudeamus.

— (2000b) "Kansalliset ja kansainväliset tietoyhteiskuntastrategiat", in M. Vuorensyrjä and R. Savolainen (eds) *Tieto ja Tietoyhteiskunta*, Helsinki: Gaudeamus.

— (ed.) (2001) *Informational Societies: Understanding the Third Industrial Revolution*, Tampere: Tampere University Press.

Kasvio, A. (2000) *Introduction*. Online. Available HTTP: http://www.info.uta.fi/winsoc/engl/lect/INTRODUCTION.html

Kautonen, M., Schienstock, G., Sjöholm, H. and Huuhka, P. (1998) *Tampereen seudun osaamisintensiiviset yrityspalvelut, Tampereen seudun osaamisintensiiviset yrityspalvelut (TOP) - projektin loppuraportti*, Tampereen yliopisto, Työelämän tutkimuskeskus, Työraportteja 56, Tampere.

Kevätsalo, K. (1999) *Jäykät joustot ja tuhlatut resurssit*, Tampere: Vastapaino.

Kinnunen, P., Leino, M., Särkelä, R. and Väärälä, R. (1998) "Verkostokoulutuksen koke-muksia" (Experiences from network education), in M. Eskelinen *et al.* (eds) *Toisin toimien uutta oppien*, Helsinki: Sosiaali- ja terveysturvan keskusliitto, 13–27.

Kivimäki, R. (1996) "Yrittäjät, perhe ja sukupuoli", in M. Kinnunen and P. Korvajärvi (eds) *Työelämän sukupuolistavat käytännöt*, Tampere: Vastapaino.

Klein, J.A. (1994) "The paradox of quality management: commitment, ownership, and control", in C. Heckscher and A. Donnellon (eds) *The Post-Bureaucratic Organization: New Perspectives on Organizational Change*, Thousand Oaks: Sage.

Kling, R. (ed.) (1996) *Computerization and Controversy*, 2nd edition, San Diego: Academic Press.

Knights, D., Murray, F. and Willmott, H. (1997) "Networking as knowledge work: a study of strategic inter-organizational development in the financial services industry", in B.P. Bloomsfield, R. Coombs, D. Knights and L. Dale (eds) *Information Technology and Organi-zations*, Oxford: Oxford University Press.

Knoblauf, H. (2001) "Fokussierte Ethnographie", *Sozialer Sinn*, 1: 123–41.

Knorr, K. (1979) "Contextuality and indexicality of organizational action: towards a tran-sorganizational theory of organizations", *Social Science Information*, 18, 1: 79–101.

Koistinen, P. and Sengenberger, W. (eds) (2002) *Labour Flexibility: A Factor of the Economic and Social Performance of Finland in the 1990s*, Tampere: Tampere University Press.

Kolehmainen, S. (2001) *Work Organisation in High-tech IT Firms*, Working Papers 62, Sitra Reports series 14, Tampere: Work Research Centre, University of Tampere.

Konttinen, E. (1997) "Professionaalinen asiantuntijatyö ja sen haasteet myöhäis-modernissa", in J. Kirjonen, P. Remes and A. Eteläpelto (eds) *Muuttuva asiantuntijuus*, Jyväskylä: University of Jyväskylä, Koulutuksen tutkimuslaitos.

Kortteinen, M. (1992) *Kunnian kenttä: Suomalainen palkkatyö kulttuurisena muotona*, Hämeen-linna: Hanki ja Jää, Karisto.

Korvajärvi, P. (1990) *Toimistotyöntekijäin yhteisöt ja muutoksen hallinta*, Work Research Centre Series T/6, Tampere: University of Tampere.

— (2002) "Miten palvelualojen työpaikat kohtaavat tietotekniikan", in R. Lavikka (ed.) *Sopeudu ja vaikuta, Työn tietoistuminen ja sukupuolen tekeminen teollisuus- ja palvelualoilla*, Tampere: Tampereen yliopisto, Työelämän tutkimuskeskus.

Korvajärvi, P. and Lavikka, R. (2000) "Diversities within the economic success, regional case study report of Finland", unpublished report of the Project Information Society, Work and the Generation of New Forms of Social Exclusion, Work Research Centre, University of Tampere, September 2000.

Kraut, R., Dumais, S. and Koch, S. (1989) "Computerization, productivity, and quality of work-life", *Communications of the ACM*, 32, 2: 220–38.

Kristeva, J. (1995) *New Maladies of the Soul: European Perspectives*, New York: Columbia University Press.

Kumar, K. (1995) *From Post-Industrial to Post-Modern Society: New Theories of the Contemporary World*, Oxford: Blackwell.

Kunda, G. (1992) *Engineering Culture: Control and Commitment in a High-tech Corporation*, Philadelphia: Temple University Press.

Kuosa, T. (2000) "Masculine world disguised as gender neutral", in E. Balka and R. Smith (eds) *Proceedings of IFIP TC9 WG9.1 Seventh International Conference on Women, Work and Computerization*, Boston: Kluwer Academic Publishers.

— (2001) "Technological determinism in IS specialists' talk", in S. Björnestad, R.E. Moe and A.I. Opdahl (eds) *Proceedings of the 24th Information Systems Research Seminar in Scandinavia (IRIS24)*, Ulvik, Norway: IRIS, 11–14.8.2001.

Kuusinen, R. (2001) *Ongelmana yhteistyökyvyttömyys? Teoreettisen ymmärryksen etsintää web-avusteiselle tiedontuottamisyhteistyölle*, Helsinki: Helsingin yliopisto, Kasvatuspsykologian tutkimusyksikkö 2/2001.

Lash, S. (1995) "Refleksiivisyys ja sen vastinparit: rakenne, estetiikka, yhteisö", in U. Beck, A. Giddens and S. Lash, *Nykyajan jäljillä*, Jyväskylä: Gummerus Kirjapaino Oy.

— (2002) *Critique of Information*, London, Thousand Oaks, New Delhi: Sage.

Lash, S. and Urry, J. (1994) *Economies of Signs and Space*, London: Sage.

Lassila, K. and Brancheau, J.C. (1999): "Adoption and utilization of commercial software packages: exploring utilization equilibria, transitions, triggers, and tracks", *Journal of Management Information Systems*, 16, 2: 63–90.

Latour, B. (1988) "Mixing humans and nonhumans together: the sociology of a door closer", *Social Problems*, 35, 3: 298–310.

— (1999) 'On recalling ANT', in J. Law and J. Hassard (eds) *Actor Network Theory and After*, Oxford: Blackwell Publishers/*The Sociological Review*.

Launis, K. (1997) "Moniammatillisuus ja rajojen ylitykset asiantuntijatyössä", in J. Kirjonen, P. Remes and A. Eteläpelto (eds) *Muuttuva asiantuntijuus*, Jyväskylä: University of Jyväskylä, Koulutuksen tutkimuslaitos.

Lave, J. and Wenger, E. (1991) *Situated Learning – Legitimate Peripheral Participation*, Cambridge: Cambridge University Press.

Lavikka, R. (1992) *Ryhmätyö tulee vaatetusteollisuuteen, Tutkimus vaatetusyritysten siirtymisestä vaihetyöstä ryhmätyöhön perustuviin työorganisaatioihin*, Working Paper 29/1992, Tampere: University of Tampere, Research Institute for Social Sciences, Work Research Centre.

— (1997) *Big Sisters, Spacing Women Workers in the Clothing Industry: A Study on Flexible Production and Flexible Women*, Publications Series T 16/1997, Tampere: University of Tampere, Research Institute for Social Sciences, Work Research Centre.

— (1998) "Yhteiseen hyvään ponnistava sisarkunta", in T. Heiskanen, R. Lavikka, L. Piispa and P. Tuuli, *Joustamisen monet muodot, Pukineteollisuus etsimässä tietä huomiseen*, Tampere: Publication Series T 17/1998, Tampere: University of Tampere, Research Institute for Social Sciences, Work Research Centre.

— (2001) *Final National Report of Tampere Region, Finland, for the European Commission's EU Targeted Socio-Economic Research Programme SOWING: Information Society, Work and the Generation of New Forms of Social Exclusion*, Tampere: University of Tampere, Research Institute for Social Sciences, Work Research Centre.

— (2002) "Johtopäätöksiä organisaatiokulttuurista, työn tietoistumisesta ja sukupuolesta teollisuus- ja palveluyrityksissä", in R. Lavikka (ed.) *Sopeudu ja vaikuta, Työn tietoistuminen ja sukupuolen tekeminen teollisuus- ja palvelualoilla*, Tampere: Tampereen yliopisto, Työelämän tutkimuskeskus.

Law, J. (1998) "Machinic pleasures and interpellations", in B. Brenna, J. Law and I. Moser (eds) *Machines, Agency and Desire*, TMW Skriftserie Nr. 33, Oslo: University of Oslo, Centre for Technology and Culture.

— (2001) "Machinic pleasures and interpellations". Available HTTP: http://www.comp.lancs.ac.uk/sociology/soc067jl.html

Law, J. and Hassard, J. (1999) *Actor Network Theory and After*, Oxford: Blackwell.

Lefebvre, H. (1998) *The Production of Space*, Oxford and Cambridge: Blackwell.

Lehtimäki, H. (1999) "Verkostopuhe alueellisen tulevaisuuden toimintatiloja tehtäessä", in P. Eriksson and M. Vehviläinen (eds) *Tietoyhteiskunta seisakkeella: teknologia, strategiat ja paikalliset tulkinnat*, Jyväskylä: Sophi.

Lehtinen, E. and Palonen, T. (1997) "Tiedon verkottuminen – haaste asiantuntijuudelle", in J. Kirjonen, P. Remes and A. Eteläpelto (eds) *Muuttuva asiantuntijuus*, Jyväskylän: Jyväskylän yliopisto, Koulutuksen tutkimuslaitos.

Lehto, A.-M. (1999) "Women in working life in Finland", in *Women in Finland*, Helsinki: Otava.

Lehto, A.-M. and Sutela, H. (1999) *Gender Equality in Working Life*, Helsinki: Statistics Finland, Labour Market, 22.

— (1999) *Tasa-arvo työoloissa*, Helsinki: Tilastokeskus.

Lehtonen, H. (1990) *Yhteisö*, Tampere: Vastapaino.

Levi-Strauss, C. (1973) *Structural Anthropology*, London: Allen Lane. Cited in S. Wright (1998) "The politicization of 'culture'". Presidential address, Section H, Association for the Advancement of Science, Royal Anthropological Institute. Available HTTP: http://lucy.ukc.ac.uk/rai/AnthToday/wright.htm

Lie, M. (1995) "Technology and masculinity: the case of computer", *European Journal of Women's Studies*, 2, 3: 379–94.

Lincoln, J.R. and Kalleberg, A.L. (1990) *Culture, Control, and Commitment: A Study of Work Organization and Work Attitudes in the United States and Japan*, Cambridge: Cambridge University Press.

Lister, R. (1997) *Citizenship: Feminist Perspectives*, London: Macmillan.

Loader, B.D. (1998) (ed.) *Cyberspace Divide: Equality, Agency and Policy in the Information Society*, London: Routledge.

Locksley, G. (1986) "Information technology and capitalist development", *Capital & Class*, 27: 81–106.

Lyytinen, K. and Goodman, S. (1999) "Finland: the unknown soldier on the information technology front", *Communications of the ACM*, 42, 3: 13–17.

Macdonald, L. (1997) "Classical social theory with the women founders included", in C. Camic (ed.) *Reclaiming the Sociological Classics*, Malden, MA: Blackwell.

McDonough, J.P. (1999) "Designer selves: construction of technologically mediated identity within graphical, multiuser virtual environments", *Journal of the American Society for Information Science*, 50, 10: 855–69.

Mackay, H., Carne, C., Beynon-Davis, P. and Tudhope, D. (2000) "Reconfiguring the user: using Rapid Application Development", *Social Studies of Science*, 30, 5: 737–57.

MacKenzie, D. and Wajcman, J. (eds) (1985) *The Social Shaping of Technology: How the Refrigerator Got Its Hum*, Milton Keynes: Open University Press.

— (eds) (1999) *The Social Shaping of Technology*, 2nd edition, Buckingham: Open University Press (1st edition published 1985).

MacKinlay, J.B. (1973) "On the professional regulation of change", in P. Halmos (ed.) *Professionalisation and Social Change*, Keele: Sociological Review Monograph 20.

McLoughlin, I. (1999) *Creative Technological Change: The Shaping of Technology and Organisations*, London and New York: Routledge.

Margolis, J. and Fisher, A. (2002) *Unlocking the Clubhouse: Women in Computing*, Cambridge, Mass.: MIT Press.

Markus, M.L. and Benjamin, R.I. (1997) "The magic bullet theory in IT-enabled transformation", *Sloan Management Review*, 38, 2: 55–68.

Markus, M.L. and Robey, D. (1988) "Information technology and organizational change: causal structure in theory and research", *Management Science*, 5, 34: 583–98.

Masuda, Y. (1980) *The Information Society as a Post-industrial Society*, Tokyo: Insitute for the Information Society.

Mead, G.H. (1976) *Medvetandet, jaget och samhället från socialbehavioristisk ståndpunkt*, Kalmar: Argos. (Licensed by the University of Chicago.)

Meyer, J.P. and Allen, N.J. (1997) *Commitment in the Workplace: Theory, Research and Application*, Thousand Oaks, CA: Sage.

Meyrowitz, J. (1985) *No Sense of Place: The Impact of Electronic Media on Social Behavior*, New York: Oxford University Press.

Miles, I. and Kastrinos, N., with Flanagan, K., Bilderbeek, R. and Den Hertog, P., with Huntik, W. and Bouman, M. (1995) *Knowledge-intensive Business Services, Users, Carriers and Sources of Innovation*, EIMS Publication No 15, European Innovation Monitoring System (EIMS).

Mintzberg, H. (1989) *Mintzberg on Management, Inside Our Strange World of Organizations*, New York: The Free Press.

Mol, A. (1999) "Ontological politics, a word and some questions", in J. Law and J. Hassard (eds) *Actor Network Theory and After*, Oxford: Blackwell.

Moore, H. (1987) *Feminism and Anthropology*, Cambridge: Cambridge University Press.

Morris, M. (1988) *The Pirate's Fiancée*, London: Verso.

Morrow, P. (1993) *The Theory and Measurement of Work Commitment*, Greenwich: JAI Press.

Mörtberg, C. (1997) "'Det beror på att man är kvinna ...' Gränsvandrerskor formas och formar informationsteknologi", doctoral dissertation, Luleå: Institutionen för Arbetsvetenskap 1997: 12, Avdelningen Genus och Teknik, Luleå Tekniska Universitet.

Moser, I. (1998) "The rehabilitation of the cyborg", manuscript, Centre for Technology and Culture, University of Oslo.

— (2000) "Against normalization: subverting norms of ability and disability", *Science as Culture*, 9, 2: 201–40.

Mumford, E. and Henshall, D. (1979) *A Participative Approach to Computer Systems Design*, London: Associated Business Press.

Nambisan, S., Agarwal, R. and Tanniru, M. (1999) "Organizational mechanisms for enhancing user innovation in information technology", *MIS Quarterly*, 23, 3: 365–95.

Ngwenyama, O.K. (1998) "Groupware, social action and organizational emergence: on the process dynamics of computer mediated distributed work", *Accounting, Management and Information Technology*, 8: 127–46.

Niemelä, J. (1996) *Ammattirajoista tiimityöskentelyyn: työnjaon ja työelämän suhteiden muutos Suomen telakoilla 1980- ja 1990-luvuilla*, Publications of the University of Turku, Ser. C /127, Turku: University of Turku.

Nonaka, I. and Konno, N. (1998) "The concept of 'ba': building a foundation for knowledge creation", *California Management Review*, 40, 3: 40–54.

Nonaka, I and Takeuchi, H. (1995) *The Knowledge-creating Company: How Japanese Companies Create the Dynamics of Innovation*, New York: Oxford University Press.

Norman, D.A. (1993) *Things that Make Us Smart: Defending Human Attributes in the Age of the Machine*, USA: Perseus Books.

Nurminen, M.I. (1986) *Kolme näkökulmaa tietotekniikkaan*, Juva: WSOY.

— (1988) *People or Computers: Three Ways of Looking at Information Systems*, Lund: Studentliteratur/Chartwell-Bratt.

O'Brien, M. and Penna, S. (1998) *Theorising Welfare*, London: Sage.

OECD (1997) *National Innovation Systems*, Paris: OECD.

Orlikowski, W.J. (1992) "The duality of technology: rethinking the concept of technology in organizations", *Organization Science*, 3, 3: 398–427.

— (1993) "CASE tools as organizational changes in systems development: investigating incremental and radical changes in systems development", *MIS Quarterly*, 17, 3: 309–40.

— (2000) "Using technology and constituting structures: a practice lens for studying technology in organizations", *Organization Science*, 11, 4: 404–28.

Orlikowski, W.J. and Gash, D.C. (1994) "Technological frames: making sense of information technology in organizations", *ACM Transactions on Information Systems*, 12, 2: 174–207.

Orlikowski, W.J. and Robey, D. (1991) "Information technology and the structuring of organizations", *Information Systems Research*, 2, 2: 143–69.

Ormala, E. (1999) "Finnish innovation policy in the European perspective", in G. Schienstock and O. Kuusi *Transformation Towards a Learning Economy, The Challenge for the Finnish Innovation System*, Sitra 213, Helsinki: Sitra.

Ortner, S. (1974) "Is female to male as nature is to culture?" in M.Z. Rosaldo and L. Lamphere (eds) *Women, Culture and Society*, Stanford, CA: Stanford University Press.

Parjo, L. (ed.) (1999) *Tiedolla tietoyhteiskuntaan II*, Helsinki: Tilastokeskus.

Pekkola, J. (1993a) *Telework – a New Touch: Flexiwork Perspectives in the European and Nordic Labour Markets*, Publication of Labour Administration No. 39, Helsinki: Ministry of Labour.

— (1993b) *Etätyön soveltaminen henkilökohtaisella, tuotanto-organisaation ja työmarkkinajärjestelmän tasolla*, Työpoliittinen tutkimus No. 47, Helsinki: Työministeriö.

— (1995) *Euroopan Unionin etätyöpolitiikka ja alueiden kehittäminen, Selonteko EU-komission etätyöprojekteista, eräistä projekteista Skotlannissa ja Ranskassa sekä suosituksia toimenpiteiksi maaseutupolitiikan neuvottelukunnan etätyön teemaryhmään*, Työpoliittinen tutkimus nro 100, Helsinki: Työministeriö.

Polanyi, M. (1958) *Personal Knowledge: Towards a Post-critical Philosophy*, Chicago: University of Chicago Press.

— (1966) *The Tacit Dimension*, London: Routledge and Kegan Paul.

Porat, M.U. (1998) "The information economy: definition and measurement", in J.W. Cortada (ed.) *The Rise of the Knowledge Worker*, Boston: Butterworth-Heinemann.

Poster, M. (1990) *The Mode of Information: Poststructuralism and Social Context*, Chicago: University of Chicago Press.

Poster, W.R. (2002) "Racialism, sexuality, and masculinity: gendering 'Global Ethnography' of the workplace", *Social Politics*, 9, 1: 124–58.

Poynter, G. and de Miranda, A. (2000) "Inequality, work and technology in the services sector", in S. Wyatt, F. Henwood, N. Miller and P. Senker (eds) *Technology and In/equality: Questioning the Information Society*, London: Routledge.

Prasad, P. (1993) "Symbolic processes in the implementation of technological change: a symbolic interactionist study of work computerization", *Academy of Management Journal*, 36, 6: 1400–29.

Pratt, M. G. and Doucet, L. (2000) "Ambivalent feelings in organizational relationships", in S. Fineman (ed.) *Emotion in Organizations*, London: Sage.

Pringle, R. (1989) *Secretaries Talk*, London: Verso.

Quality of Life, Knowledge and Competitiveness: Premises and Objectives for Strategic Development of the Finnish Information Society (1998) Helsinki: Finnish National Fund for Research and Development, Sitra.

Qvortrup, L. (1998) "From teleworking to networking: definitions and trends", in P.J. Jackson and J.M. van der Wielen (eds) *Teleworking: International Perspectives, from Telecommuting to the Virtual Organisation*, London: Routledge.

Rantalaiho, L. (1997) "Contextualizing gender", in L. Rantalaiho and T. Heiskanen (eds) *Gendered Practices in Working Life*, London and New York: Macmillan and St Martin's Press.

Rantalaiho, L. and Heiskanen, T. (eds) (1997) *Gendered Practices in Working Life*, London and New York: Macmillan Press and St Martin's Press.

Rantanen, J. (2000) "Tietointensiivisen työn kehitysnäkymiä Suomessa", *Työ ja Ihminen*, 14, 2: 89–93.

Reich, R.B. (1992) *The Works of Nations*, New York: Vintage Books.

Resnick, L.B. (1996) "Situated rationalism: the biological and cultural foundations for learning", *Prospects*, 26, 1: 37–53.

Rheingold, H. (1995) *The Virtual Community, Finding Connection in a Computerized World*, London: Minerva.

Ricci, A. (2000): "Measuring information society: dynamics of European data on usage and communication technologies in Europe since 1995", *Telematics and Informatics*, 17, 1–2: 141–67.

Riggins, S.H. (1997) "The rhetoric of othering", in S.H. Riggins (ed.) *The Language and Politics of Exclusion: Others in Discourse*, Thousand Oaks, CA: Sage.

Robey, D. (1997) "The paradoxes of transformation", in C. Sauer and P.W. Yetton (eds) *Steps to the Future: Fresh Thinking on the Management of IT-based Organizational Transformation*, San Francisco: Jossey-Bass Publishers.

Rockart, J.F. and Flannery, L.S. (1983) "The management of end user computing", *Communications of the ACM*, 26, 10: 776–84.

Roethlisberger, F.J. and Dickson, W.J. (1949) *Management and the Worker*, Cambridge, MA: Harvard University Press.

Rommes, E. (2000) "Gendered user-representations, design of a digital city", in E. Balka and R. Smith (eds) *Proceedings of IFIP TC9 WG9.1 Seventh International Conference on Women, Work and Computerization*, Boston: Kluwer Academic Publishers.

Roszak, T. (1986) *The Cult of Information: The Folklore of Computers and the True Art of Thinking*, Cambridge: Lutterworth.

Ryle, G. (1949/1963) *The Concept of Mind*, Harmondsworth: Penguin.

Said, E. (1978) *Orientalism*, Harmondsworth: Penguin.

Salmi, M. (1991) *Ansiotyö kotona – toiveuni vai painajainen*, Helsingin yliopiston laitoksen tutkimuksia No 225, Helsinki: Yliopistopaino.

Sauer, C. and Yetton, P.W. (eds) (1997): *Steps to the Future: Fresh Thinking on the Managemennt of IT-based Organizational Transformation*, San Francisco: Jossey-Bass.

Scarbrough, H. (1993) "Problem-solutions in the management of information systems expertise", *Journal of Management Studies*, 30, 6: 939–55.

Schienstock, G. (1997) "Information society, work and the generation of new forms of social exclusion: a research plan for the TSER-programme of the European Union", unpublished.

Schienstock, G.., Bechmann, G., Flecker, J., Huws, U., Van Hootegem, G., Mirabile, M., Moniz, A., and Ó Siochru, S. (1999) *Information Society, Work and the Generation of New Forms of Social Exclusion (SOWING)*, First Interim Report (Literature Review), Tampere, Finland. Available HTTP: http://www.uta.fi/laitokset/tyoelama/sowing/frontpage.html

Schienstock, G., Rissanen, T. and Timonen, H. (2001) *Pirkanmaalaiset yritykset matkalla tieto-yhteiskuntaan, Yritysten teknologiset käytännöt eurooppalaisessa vertailussa*, Working Papers

63/2001, Tampere: University of Tampere. Research Institute for Social Sciences, Work Research Centre.

Schiller, D. (1999)*Digital Capitalism: Networking the Global Market System*, Cambridge, Mass.: MIT Press.

Schiller, H.I. (1976) *Communication and Cultural Domination*, New York: International Arts and Sciences Press.

Sennett, R. (1998) *The Corrosion of Character: The Personal Consequences of Work in the New Capitalism*, New York, London: W.W. Norton.

—— (2000) "Street and office: two sources of identity", in W. Hutton and A. Giddens (eds*)* *On the Edge: Living with Global Capitalism*, London: Jonathan Cape.

Shields, R. (2002) *The Virtual*, London: Routledge.

Smith, D.E.(1990) *The Conceptual Practices of Power: A Feminist Sociology of Knowledge*, Toronto: University of Toronto Press.

Socially Responsible Information Society: Finland's New International Identity (2001) Available HTTP: http://www.tt.fi/arkisto/gethtml.pl?ft_cid=2116

Soja, E.W. (1996*) Thirdspace, Journeys to Los Angeles and Other Real-and-imagined Places*, Oxford: Blackwell.

—— (2000) *Postmetropolis: Critical Studies of Cities and Regions*, Oxford: Blackwell.

Sorokin, P. (1943) *Sociocultural Causality, Space, Time*, Durham, NC: Duke University Press.

Spender, J.C. (1996) "Making knowledge the basis of a dynamic theory of the firm", *Strategic Management Journal*,17: 45–62.

Stacey, M. (1974) "The myth of community studies", in C. Bell and H. Newby (eds*) The Sociology of Community*, London: Frank Cass, 13–26.

Ståhle, P. and Grönroos M. (1999) *Knowledge Management – tietopääoma yrityksen kilpailuetuna*, Helsinki: WSOY.

Stanworth, C. (1998) "Telework and information age", *New Technology, Work, and Employment*,13, 1: 51–62.

Star, S.L. (1991) "Power, technology and the phenomenology of conventions: on being allergic to onions", in J. Law (ed.*) A Sociology of Monsters? Power, Technology and the Modern World*, Oxford: Blackwell.

—— (1999) "The ethnography of infrastructure", *American Behavioral Scientist*, 43, 3: 377–91.

Starbuck, W.H. (1992) "Learning by knowledge-intensive firms", *Journal of Management Studies*, 29, 6: 713–40.

Statistics Finland (2002) *Finland in Figures 2002*, Helsinki: Statistics Finland.

Stewart, T.A. (1997) *Intellectual Capital: The New Wealth of Organizations*, New York: Currency Doubleday.

Strathern, M. (2002) "Abstraction and decontextualization: an anthropological comment", in S. Woolgar (ed.) *Virtual Society? Technology, Cyberbole, Reality*, Oxford: Oxford University Press.

Suomi tietoyhteiskunnaksi (1995) Kansalliset linjaukset,Tikas-ohjausryhmän loppuraportti, Vantaa: Painatuskeskus Oy.

Sveiby, K.E. (1996) *The Knowledge Organisation*. Online. Available HTTP: http://www.sveiby.com.au/KOS1.html

Sveiby, K.E. and Risling, A. (1987) *Tietoyrityksen johtaminen-vuosisadan haaste?* Espoo: Weiling+Göös.

Swingewood, A. (1977) *The Myth of Mass Culture*, London: Macmillan.

Tähkä, V. (1996) *Mielen rakentuminen ja psykoanalyyttinen hoitaminen*, Juva: WSOY.

Tancred-Sheriff, P. (1989) "Gender, sexuality and the labour process", in J. Hearn, D.L. Sheppard, P. Tancred-Sheriff and G. Burrell (eds) *The Sexuality of Organization*, London: Sage.

Tapscott, D. (1998) *Growing Up Digital: The Rise of the Net Generation*, New York: McGraw-Hill.

Taylor, C. (1989) *Sources of the Self, the Making of the Modern Identity*, New York: Cambridge University Press.

—— (1995) *Autenttisuuden etiikka*, Helsinki: Gaudeamus.

Taylor, S. (1998) "Emotional labour and the new work place", in P. Thompson and C. Warhust (eds) *Workplace of the Future*, London: Macmillan.

Taylor, S. and Tyler, M. (2000) "Emotional labour and sexual difference in the airline industry", *Work, Employment and Society*, 14, 1: 77–95.

Thompson, P. and McHugh, D. (1990) *Work Organisations: A Critical Introduction*, London: Macmillan.

Tidd, J., Bessant, J. and Pavitt, K. (1997) *Managing Innovation: Integrating Technological, Market and Organisational Change*, Chichester: John Wiley.

Tilastokeskus (2002) *Suomi lukuina*, Helsinki: Tilastokeskus.

Toffler, A. (1980) *The Third Wave*, London: Pan Books.

Tolsby, J. (2000) "Taylorism given a helping hand: how IT systems changed employees' flexibility and personal involvement in their work", *Journal of Organizational Change Management*, 13, 5: 482–92.

Tönnies, F. (1967) *Community and Society*, Lansing: Michigan State University Press.

Truex, D.P., Baskerville, R. and Klein, H. (1999) "Growing systems in emergent organizations", *Communications of the ACM*, 42, 8: 117–23.

Tuomi, I. (1999) *Corporate Knowledge: Theory and Practice of Intelligent Organisations*, Helsinki: Metaxis.

Turoff, M. (1997) "Virtuality", *Communications of the ACM*, 40, 9: 38–43.

Tuuva, S. (2000) "Sinuja koneen kanssa – Tulkintoja tietotekniikasta" [Well acquainted with the machine – interpretations of information technology], *Tiedepolitiikka* 3/2000.

Tuuva, S. and Uotinen, J. (1999) "Tiedon valtateiltä kinttupoluille" [From the information highways to paths], in *Tietoyhteiskunta seisakkeella: teknologia, strategiat ja paikalliset tulkinnat*, SoPhi, Jyväskylä, 1999, 203–14.

Uotinen, J., Tuuva, S., Vehviläinen, M. and Knuutila, S. (eds) (2001) *Verkkojen kokijat paikallista tietoyhteiskuntaa tekemässä*, Saarijärvi: Gummeruksen Kirjapaino Oy.

Urry, J. (1985) "Social relations, space and time", in D. Gregory and J. Urry (eds) *Social Relations and Spatial Structures*, New York: St Martin's Press.

Van Maanen, J. (1988) *Tales of the Field: On Writing Ethnography*, Chicago: University of Chicago Press.

Vehviläinen, M. (1997) *Gender, Expertise and Information Technology*, doctoral dissertation, Tampere: University of Tampere, Department of Computer Science A–1997–1.

— (1999a) "Naisten tietotekniikkaryhmä: yhteisöllisestä ja paikallisesta kansalaisuudesta", in P. Eriksson and M. Vehviläinen (eds) *Tietoyhteiskunta seisakkeella: Teknologia, strategiat ja paikalliset tulkinnat*, Jyväskylä: Sophi.

— (1999b) "Gender and computing in retrospect: the case of Finland", *IEEE Annals of the History of Computing*, 21, 2: 44–51.

— (2000) "Gender and information technology", in C. Mörtberg (ed.) *Where Do We Go from Here? Feminist Challenges of Information Technology*, Division Gender and Technology, Luleå: Luleå University of Technology.

— (2001a) "Gender and citizenship in the information society: women's information technology groups in North Karelia", in A. Adam and E. Green (eds) *Virtual Gender*, London: Routledge.

— (2001b) "Gendered technological nationalism", paper at the 5th Conference of the European Sociological Association "Visions and Divisions", 28 August to 1 September 2001, Helsinki.

— (2001c) "Paikallista tietoyhteiskuntaa tekemässä", in J. Uotinen, S. Tuuva, M. Vehviläinen and S. Knuutila (eds) *Verkkojen kokijat paikallista tietoyhteiskuntaa tekemässä*, Saarijärvi: Gummeruksen Kirjapaino Oy.

— (2002) "Gendered agency in information society: on located politics of technology", in M. Consalvo and S. Paasonen (eds) *Women and Everyday Uses of the Internet: Agency and Identity*, New York: Peter Lang.

— and Eriksson, P. (1999) "Teknologia, strategiat ja paikalliset tulkinnat", in P. Eriksson and M. Vehviläinen (eds*) Tietoyhteiskunta seisakkeella: teknologia, strategiat ja paikalliset tulkinnat*, Jyväskylä: Sophi.

Virkkunen, J. (1990*) Johtamisen rationalisointi vai kehityksen hallinta: tulosjohtamisen tehokkuuskäsitys ja sen ylittämisen mahdollisuudet*, Helsinki: Julkishallinnon kouluttajat ry.

Waddington, J. (ed.) (1999) *Globalization and Patterns of Labour Resistance*, London: Mansell.

Wajcman, J. (1991) *Feminism Confronts Technology*, Cambridge: Polity Press.

Walton, R.E. (1980) "Establishing and maintaining high commitment work systems", in J.R. Kimberly and R.H. Miles (eds) *The Organizational Life Cycle*, San Francisco: Jossey-Bass.

— (1985) "From control to commitment in the workplace", *Harvard Business Review*, 64, 2: 77–84.

Wang, G. (1994) *Treading Different Paths: Informatization in Asian Nations*, Norwood, NJ: Ablex.

Watson-Manheim, M.B., Chudoba, K.M. and Crowston, K. (2002) "Discontinuities and continuities: a new way to understand virtual work", *Information Technology & People*, 15, 3: 191–209.

Webster, F. (1995) *Theories of the Information Society*, London: Routledge.

— (2002) *Theories of the Information Society*, 2nd edition, London: Routledge.

Webster, J. (1996) *Shaping Women's Work: Gender, Employment and Information Technology*, London: Longman.

Webster, J. and Martocchio, J.J. (1992) "Microcomputer playfulness: development of a measure with workplace implications", *MIS Quarterly*, 16, 2: 201–26.

Weick, K.E. (1995) *Sensemaking in Organizations*, Thousand Oaks: Sage.

Weiler, K. (1988) *Women Teaching for Change: Gender, Class and Power*, New York: Bergin & Garvey.

Wenger, E.C. (1998) *Communities of Practice – Learning, Meaning, and Identity*, Cambridge: Cambridge University Press.

Wenger, E.C. and Snyder, W.M. (2000) "Communities of practice: the organizational frontier", *Harvard Business Review*, 78, 1: 139–45.

Whyte, W.H. (1956) *The Organization Man*, New York: Doubleday Anchor.

Wilenius, M. and Kamppinen, M. (2000) "Living beyond the information society", *Foresight*, 2, 2: 147–50.

Wilensky, H.L. (1967) *Organizational Intelligence: Knowledge and Policy in Government and Industry*, New York: Basic Books.

Williams, R. (1976) *Keywords*, Glasgow: Fontana.

Winner, L. (1993) "Social constructivism: opening the black box and finding it empty", *Science as Culture*, 3, 3: 427–52.

Winslow, C.D. and Bramer, W.L. (1994) *FutureWork: Putting Knowledge to Work in the Knowledge Economy*, New York: Free Press.

Wise, J.M. (1997) *Exploring Technology and Space*, Thousand Oaks: Sage.

Woolgar, S. (ed.) (2002) *Virtual Society? Technology, Cyberbole, Reality*, Oxford: Oxford University Press.

World Employment Report 2001 Life at Work in the Information Economy (2001) Geneva: International Labour Organization. Available HTTP: http://www.ilo.org/public/english/support/publ/wer/overview.htm

Wright, S. (1998) "The politicization of 'culture'". Presidential address, Section H, Association for the Advancement of Science, Royal Anthropological Institute. Available HTTP: http://lucy.ukc.ac.uk/rai/AnthToday/wright.htm

Wyatt, S. (2001) "The information society: what it means to be out of the loop", in E. Pantzar, R. Savolainen and P. Tynjälä (eds) *In Search of a Human-centred Information Society*, Tampere: Tampere University Press.

Wyatt, S., Henwood, F., Miller, N. and Senker, P. (eds) (2000) *Technology and In/equality: Questioning the Information Society*, London: Routledge.

Ylöstalo, P. and Kauppinen, T. (1995) *Työolobarometri marraskuu 1994*, Työpoliittinen tutkimus nro 112, Helsinki: Työministeriö.

Zuboff, S. (1988) *In the Age of the Smart Machine: The Future of Work and Power*, New York: Basic Books.

Index